Karl Sommer

Sampling of Powders and Bulk Materials

With 45 Figures

Springer-Verlag
Berlin Heidelberg New York Tokyo

Prof. Dr.-Ing. Karl Sommer
Lehrstuhl für Maschinen- und Apparatekunde
Technische Universität München
8050 Freising-Weihenstephan, FRG

Library of Congress Cataloging in Publication Data
Sommer, Karl
Sampling of powders and bulk materials.
Bibliography: p.
Includes index.
1. Bulk solids--Sampling. 2. Powders--Sampling. 3. Sampling. I. Title.
TS180.8.B8S6 1985 660.2'82'0287 85-22140

ISBN-13: 978-3-642-82607-8 e-ISBN-13: 978-3-642-82605-4
DOI: 10.1007/978-3-642-82605-4

Softcover reprint of the hardcover 1st edition 1986

2161/3020-543210

Preface

The present book is based on the experience of sampling powders and bulk materials at institute of "Mechanische Verfahrenstechnik", Karlsruhe University (Germany).

Sampling is understood to be the taking of samples from a material in order to obtain information about the composition of the total amount in respect of one or more attributes, such as metal content, ash content, moisture, particle size, particle shape etc. In the case of granular bulk material already comminuted or not, samples from tubs, trucks, bags or barrels. As granular material has a strong tendency to segregate, portions must be removed form these containers at many places.

The book describes the basic theory behind sampling materials in powder or granular form. In addition to providing fundamental statistical data, it also highlights the mistakes that can arise in the different types of sampling techniques, such as suspensions, pellet cross-sections, sample dividing, or sampling according to number or weight.

If the samples contain individual particles, comparable in size to the size of the sample itself then in addition to the number of individual elements in the sample and the type of sampling the particle size and particle size distribution will be further variables. This book will pay special attention to this fact important for the whole of powder technology, the chemical, pharmaceutical, food and and for mineral processing industry.

The present theory seems to be the first which gives results for all common types of sampling e.g. numerical or mass concentration by constant particle numbers, constant sample mass or sample divider.

Types of sampling usual in praxis will be presented as tables, in form of "cookery formulas" to make the common work easy.

Weihenstephan, December 1985 Karl Sommer

Contents

1 Introduction

Sampling is understood to be the taking of samples from a material in order to obtain information about the composition of the total amount in respect of one or more attributes, such as metal content, ash content, moisture, particle size, particle shape etc. The sample is a portion of the total material under consideration and is always resorted to in the investigation of the attributes, whenever a complete, ie. 100 % examination of the total material is not possible, be it for economic reasons, when the method required for the investigation of large total amounts cannot be justified or when the product is destroyed during the analysis and can no longer be used.

Sampling is carried out with the aid of a sampling scheme in which the type of sampling is laid down. Accordingly, the selection is carried out systematically or randomly. In the case of a random sample, the individual samples, also known as elements, are taken independently of one another and randomly from the total material under investigation. All individual samples of the same size must have the same probability of being selected into the total sample. In contrast to the random sample, the systematic sample consists of individual samples drawn at regular intervals of time, quantity or space, distributed over the total amount of material. The first individual sample is still selected randomly. When choosing the interval between two individual samples, care must be taken that the sequence of the individual samples does not coincide with the natural variations of the attributes, as this can cause unforeseeable sampling errors. Despite these possible sources of error, the regular distribution of individual samples, that is systematic sampling, is preferred, because the administration is simpler. The sampling scheme also provides information about the number of the individual samples contained in the sample, which depends on the precision required, ie. on the variation, to be expected within defined confidence limits, in the results of the investigation. The variations in the results of the investigation in respect of the true content in the total amount are called sampling errors.

These errors are systematic errors if they lead to discrepancies despite theoretically infinitely frequent repetition of the investigation. These systematic errors are process-dependent and are not the subject of the present paper. On the other hand, though random errors do lead to uncertainties, on average the true value can be expected as the measured value. The determination of these errors is a task for mathematical statistics. The most important parameters to be investigated are the variances and their roots, the standard deviations.

The most important variable, apart from the type of sampling, is the sample size. It determines to a great extent the size of the variations to be expected. Conversely, a minimum number of individual samples corresponds to a claim to precision with a given statistical certainty (confidence region). In the case of ideally homogeneous material, one individual sample is theoretically sufficient for the determination of the content of active ingredient. In practice, however, a certain number of values of the attribute have to be determined, because of the analytical error always present. Should the material itself be inhomogeneous, a further uncertainty of measurement arises, the sampling error. The value of the attribute found with the aid of the sample will on average be the closer to the true value, the larger the sample and the more homogeneous the material. The sampling accuracy can be chekked by two-fold (or several-fold) sampling, eg. by supplier and customer.

If the material to be investigated consists of different parts, then these parts can be considered as strata of the total entity. The size and number of these strata is determined by temporal, quantitative or spatial subdivision which is appropriate, ie. will lead to greater accuracy. One or several, random or systematic individual samples are to be taken from each stratum. The number of these individual elements depends on the size and the degree of homogeneity of the individual strata. This procedure is referred to as stratified random sampling. This can lead to considerable improvement in sampling accuracy in certain cases.

When the bulk is sub-divided into stages, eg. into railway trucks, pallets, sacks or bags, only a part of the elements being selected for investigation at each stage, then the procedure is referred to as stage sampling. As is frequently the case in practice, individual stages can also come about by repeatedly dividing the sample at each selection of individual portions. The degree of homogeneity in the individual batches determines the optimum sub division of the elements to be chosen.

The choice of individual samples can be predetermined, as the example of stage sampling shows, by the method of packaging and transport. Most frequently, however, sampling of the bulk has to be carried out by hand or with the aid of a sampling tool. The technical details are frequently laid down in norms and guidelines in order to obtain the highest possible comparability of experimental results. The possible tools are of course largely dependent upon the state of the material to be investigated.

In the investigation of gases and of dustcontaining gas suspensions, partial streams are often diverted from the gas stream and measured directly or are passed on as a partial volume of gas to the laboratory for investigation. In the indirect method volumes of gas of any desired size are taken out of the gas stream with the aid of sample containers. When the pressure is sufficiently high this is most frequently possible without difficulty, otherwise the sample is taken by means of a suction device. There is always the danger in the case of the dust-laden gas sample that systematic errors arise due to the forces of fluid flow and the inertia forces having a selective effect; this must be avoided if possible.

There are similar difficulties when sampling from liquids, suspensions and slurries. For this purpose particularly pipettes and dipping flasks are used in laboratories.

When the viscosity is sufficiently high, in the case of creams, pastes or fats, the problem of selective segregation does not arise during sampling. A socalled sampler that includes the content across the whole cross-section is used to remove the sample. When the homogeneity is high, as is frequently the case, the removal of a surface sample with the aid of a spatula is often sufficient. However, the homogeneity required must be known from experience or must be determined in a prior investigation.

When the material is lumpy, such as coarse coal, lumps of iron ore or bars of metal, a certain number of individual pieces is usually selected. Frequently the samples do not contain the theoretically required, uniform quantity of the elements to be investigated, but for practical reasons this discrepancy between theory and practice is accepted. When the material is brittle further pieces are chopped off, swarf from sawing, milling or drilling being used in the case of metals. Here again systematic errors can arise, eg. by selective size reduction. If it is economically feasible, it is therefore better to comminute and mix the lumps of material selected as the sample and then prepare laboratory and analytical samples with the aid of sample dividers.

In the case of granular bulk material already comminuted, samples are usually drawn from the conveyor belt while being transported or unloaded. In the case of material already in tubs, trucks, bags or barrels, each n-th element is selected randomly or systematically for sample investigation. As granular material has a strong tendency to segregate, portions must be removed from these containers at as many places as possible over the total depth, eg. using a sampler. Further treatment is similar to that of lumpy material and is carried out by sample division until laboratory and analysis samples are obtained.

If the samples contain individual particles, comparable in size to the size of the sample itself[1)], as just mentioned above in the examples of lumpy and granular material, then in addition to the number of individual elements in the sample and the type of sampling, the particle size and the particle size distribution will be further variables. This paper will pay special attention to this fact, important for the whole of powder technology and for the mineral processing industry.

Inferences, based on the analytical values obtained, about the compositon of the bulk material cannot be drawn determinatively, but are only possible with the aid of statistics. It would therefore appear to make sense to first of all look at the statistical bases required for the determination of these relationships in sampling. In this section the statistical terms required in sampling will be defined and explained. This matter is worth stressing since frequently these terms are used ambiguously or even wrongly in the literature.

2 Fundamentals of Statistics

2.1 Frequency distribution of discrete attributes

The basis of our scientific reasoning rests on classical physics taught in school and taught to engineers also during their studies. In classical physics it is assumed that processes occur in a determinate manner, ie. given data concerning the initial state, the final state can be calculated unambiguously. However, sooner or later every scientist is forced to recognize that many processes are not fully determinable. It does not actually matter whether this indeterminacy is founded in nature, as eg. in quantum physics or whether the causes are indeterminable in practice, such as the fluctuations in measurements in experiments and in sampling. As the number of individual results increases to a certain level, it is not too difficult to recognize that even these indeterminable processes have regularities which, though not valid for the individual result, hold for the total collection as such. However, suitable observational material has to be used. This process, called descriptive statistics, was for a long time the main content of statistics. The object is to collect observational material, organize it, and represent it graphically or using formulae. This type of statistics still finds expression nowadays in the population and economic statistics published in the form of year books. The mathematical tools necessary in order to progress from descriptive statistics to modern mathematical statistics resulted from the fundamental work of Laplace, von Mises, Gauss, Student and others and from the foundation of the axiomatically based theory of probability by Kolmogoroff[2].

The methods used are independent of the area of application in each case and are based on the theory of probability already mentioned, on the one hand, and on the other on the treatment of collectives, often also known as populations.

A population consists of elements or individuals. Each of these elements of the population has a certain property or attribute. If several elements are selected from the population, then this is called a random sample or simply a sample.

Each element has a certain property, a certain attribute - that is to say, the variable assumes a certain value. These certain values are also called realizations. Often a formal distinction is made in the literature between an attribute, such as, for example, the color, the size and the like, and their realizations, such as, for example red, blue, 1,75 m or 1,85 m. The attribute as such is indicated with capital letters A or B, the realizations are indicated with a small letter x or y. This distinction cannot always be rigorously maintained; in the text that follows it will not be used.

An element can have one or several attributes (variables). Accordingly there are problems of one or several dimensions.

The first step in gaining an overall view of the set of realizations is to place them (if possible) in a sequence. This is often a sufficiently clear representation for small numbers of elements. However, for a large number of elements nothing much is to be gained. Frequently the realizations of different elements will be identical. If these elements are summarized numerically, then the so-called discrete distribution of the absolute frequency k_i is obtained.

For example, the following realizations where obtained for 20 throws of a die:

$x'_1=6, x'_2=3, x'_3=3, x'_4=1, x'_5=6, x'_6=2, x'_7=6, x'_8=5, x'_9=2, x'_{10}=6,$
$x'_{11}=4, x'_{12}=4, x'_{13}=6, x'_{14}=3, x'_{15}=4, x'_{16}=4, x'_{17}=5, x'_{18}=2, x'_{19}=5, x'_{20}=3.$

The dashed quantities mean that the realizations have not as yet been arranged in order. Now the realizations are placed in order and identical realizations are collected together. The realization $x_1 = 1$ occurred once, $x_2 = 2$ three times, $x_3 = 3$ and $x_4 = 4$ four times each, $x_5 = 5$ three times and $x_6 = 6$ five times. In the graphical representation (Figure 2.1.1) the realizations arranged in order are marked off on the abscissa and the frequency of each realization is marked off on the ordinate.

If there are k_i elements with the realization x_i in the set of n' elements, then the quotient

$$h(x_i) = \frac{\text{number of elements } x_i}{\text{total number of elements}} = \frac{k_i}{n'} \qquad (2.1.1)$$

is called the "relative frequency" of the realization x_i. It is always true that

$$0 \leq h(x_i) \leq 1 \qquad (2.1.2)$$

In Figure 2.1.1 the relative frequency has been marked off in addition to the absolute frequency as ordinate scale.

2.2 Cumulative frequency of discrete attributes

The set of all elements which in the arrangement according to order are less than or equal to a given realization is called the absolute cumulative frequency. In the graphical representation (Figure 2.1.2) the realizations arranged according to order are again marked off on the abscissa, the sum of all k_j for which $x_j \leq x_i$ is marked off on the ordinate.

If the Σk_j is again referred to the total n', then the relative cumulative frequency is obtained.

$$H(x_i) = \frac{\sum_{j=1}^{i} k_j}{n'} = \sum_{j=1}^{i} h_j \qquad (2.2.1)$$

It is true for the sum of all relative frequencies that

$$H(x_{max}) = 1 \qquad (2.2.2)$$

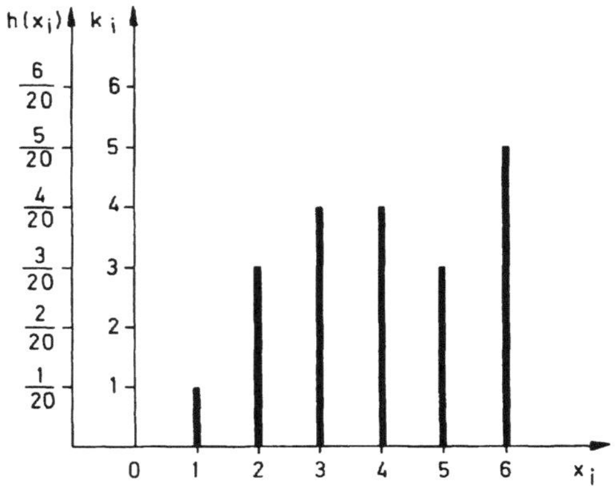

Figure 2.1.1 Absolute and relative frequency distribution of the realizations of n′ = 20 throws

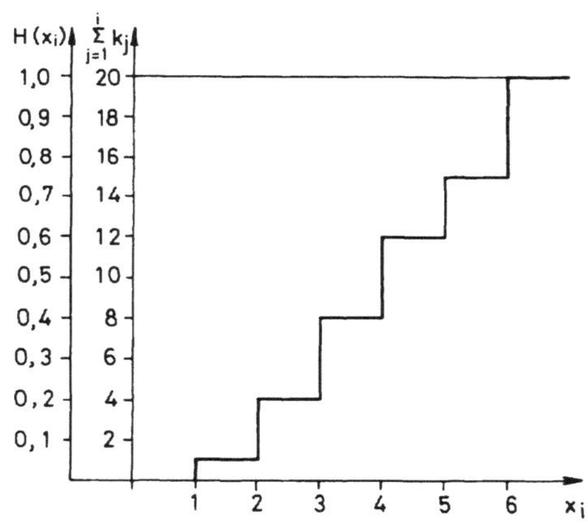

Figure 2.1.2 Absolute und relative cumulative frequency distribution of the realizations of n′ = 20 throws

In Figure 2.1.2 the relative cumulative frequency has been marked off as ordinate scale in addition to the absolute cumulative frequency.

2.3 Arithmetic mean and sample variance for discrete distributions

Mean and variance play a decisive role in statistics. Each finite collection of numbers has both these parameters. Such a set of numbers can form a distribution (Figure 2.1.1) and this distribution in turn can describe a population or a sample or a number of samples collected together. The source of these numbers does not matter as regards the definition of the mean and of the variance. If $x_1', x_2', x_3' \ldots x_n'$, and n_n' are realizations of the arbitrary attribute x, then the (arithmetic) mean is defined by

$$\bar{x} = \frac{1}{n'} \cdot \sum_{j=1}^{n'} x_j' \qquad (2.3.1)$$

and the sample variance is defined by

$$s^2 = \frac{1}{n'-1} \cdot \sum_{j=1}^{n'} (x_j' - \bar{x})^2 \qquad (2.3.2)$$

If the arithmetic mean μ of the population is known, then the sample variance can also be written:

$$s^2 = \frac{1}{n'} \cdot \sum_{j=1}^{n'} (x_j' - \mu)^2 \qquad (2.3.3)$$

The variance is thus identical to the arithmetic mean of all quadratic deviations from the mean μ.

The root of the sample variance $\sqrt{s^2} = s$ is called the sample standard deviation or "scatter". If this value s is expressed relative to the arithmetic mean $\bar{x}$, then this expression is called the sample coefficient of variation, which is given as a percentage or as a decimal value.

The ordered representation of elements in a fequency or cumulative frequency diagram contains all the information from the attributes of the elements in a clear manner. However, frequently this amount of information is too large. It may be desired to compare distributions which are not identical. Therefore the arithmetic mean is used as a parameter to characterize the location (measure of location) and the variance is used as a parameter to characterize the width of the distribution (measure of dispersion).

Normally not all the elements are different from one another in a distribution, that is to say the number of given realizations n is smaller than the total number of elements n'. If the relative frequency h_i is assigned to the elements with the realization x_i, then the arithmetic mean from equation (2.3.1) is calculated to be

$$\bar{x} = \sum_{j=1}^{n} x_j \cdot h_j \qquad (2.3.4)$$

and the sample variance from equation (2.3.3) is given by

$$s^2 = \sum_{j=1}^{n} (x_j - \mu)^2 \cdot h_j \qquad (2.3.5)$$

2.4 Frequency distribution of continuous attributes

Distributions of populations or samples are always discrete in practice. However, at least partially continuous curves instead of discrete step functions are usually employed in the mathematical treatment. Continuous distributions require a considerably smaller mathematical effort and can usually be represented with sufficient accuracy by approximation functions.

When integrating the approximation functions, non-existent integrals may arise, if the limits range from - oo to + oo. However, this only demonstrates that the approximation function does not agree sufficiently well with reality in between those limits to be able to form integrals. Since we are always dealing with finite real systems when sampling from systems of particles, the existence of the integrals of the appropriate approximation functions will not be investigated.

The approximation improves with the number of the steps of the attribute. In this case the number of elements with a certain realization is small, ie. the relative frequency of the elements with this realization is smaller, the smaller the differences between the steps of the attribute. With the transition to a continuous attribute the frequency of the elements of a certain attribute will be arbitrary. This means that a relative frequency distribution does not exist for continuous variables. However, this is not true for the relative cumulative frequency, because here in each case infinitely many, infinitely small frequencies are added. The next example will illustrate the limiting process.

An aggregate of 100,000 particles is assumed to form the frequency and cumulative frequency distributions shown in Figure 2.4.1. Initially it will be assumed that measuremens are only possible in steps of 1 millimeter. If the attribute is now measured in steps of a 1/4 millimeter, then the number found for a hitherto 1 millimeter large region is divided into four individual regions, which all have a smaller frequency. If the precision of measurement could be increased to an extreme degree, then it would be expected that each particle would have its own class. The frequeny with which elements of a class appear is then equal to the reciprocal of the given particle number, ie. extremely small.

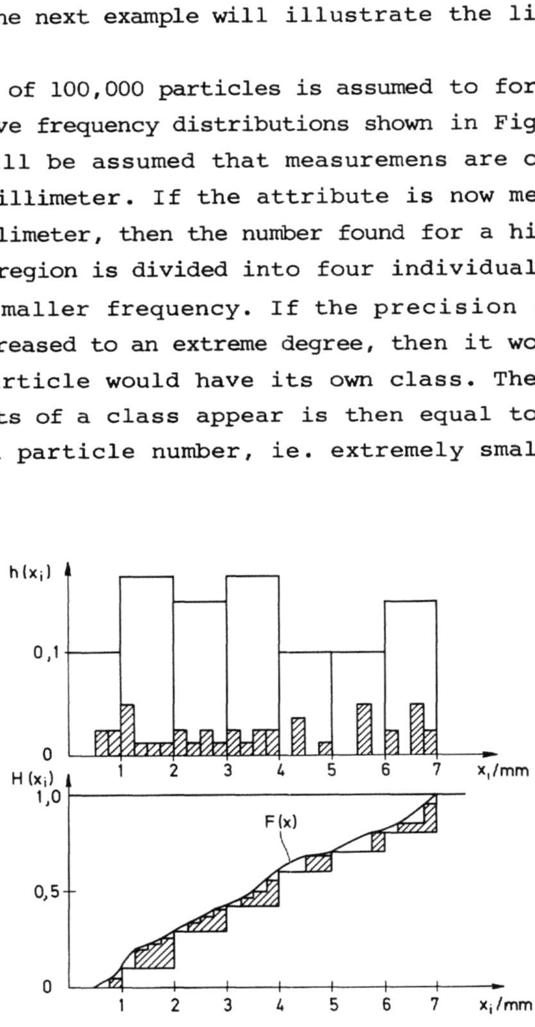

Figure 2.4.1 Limiting process of the transition from discrete frequency distributions to the continuous gradation of an attribute

However, the number of particles which are smaller than or equal to a given size does not change as the method of measurement is improved, only the gradations become smaller. At the limit, where the gradations are infinitely fine, the staircase curve becomes a smooth, uniformly increasing function F(x). The cumulative frequency thus retains its meaning as the number of all elements with an attribute that is smaller than or equal to a given attribute, even for continuous variables.

If F(x) is a sufficiently smooth function, then the differential of the distribution function F(x) with respect to the attribute x exists

$$f(x) = \frac{d\,F(x)}{dx} \qquad (2.4.1)$$

f(x) is called the distribution density funtion. The converse also holds

$$F(x) = \int_{x_{min}}^{x} (\xi)\,d\xi \qquad (2.4.2)$$

The number of those elements, whose realizations lie between two values a and b, is equal to the difference of the cumulative function between a and b and equal to the area under the distribution density f(x) (Figure 2.4.2)

$$F(a) - F(b) = \int_{a}^{b} f(x)\,dx \qquad (2.4.3)$$

The arithmetic mean μ of all possible realizations is a measure of location, just as was the case with the discrete distributions. At the limit of infinitely small steps of the attribute, the summation of equation (2.3.4) becomes an integration and then:

$$\mu = \int_{x_{min}}^{x_{max}} x\,f(x)\,dx \qquad (2.4.4)$$

Correspondingly, the variance σ^2 is given by

$$\sigma^2 = \int_{x_{min}}^{x_{max}} (x-\mu)^2 \cdot f(x)\, dx \qquad (2.4.5)$$

The root of the variance $\sqrt{\sigma^2}$ is known as the "standard deviation" or "scatter". If this value is expressed in terms of the arithmetic mean μ, then the quantity σ/μ is called the coefficient of variation. It is given as a percentage or as a decimal value.

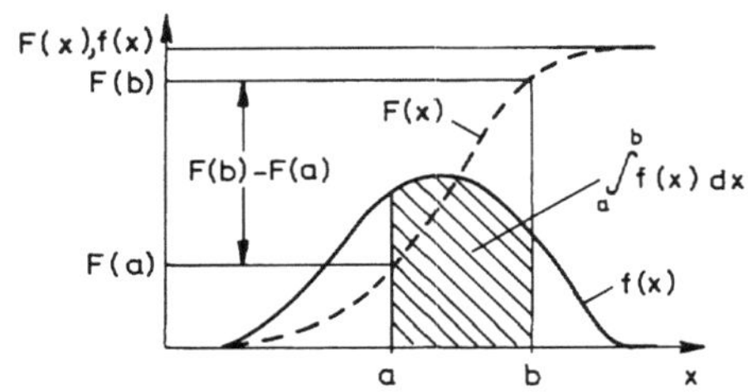

Figure 2.4.2 Diagrammatic representation of the distribution function F(x) and the distribution density f(x)

2.5 Expectation and variance of one-dimensional probability functions

If the elements with the attribute x are the result of a random process, then x is also called random variable. Thus, every sufficiently regular function g(x) is also a random variable. The distribution density f(x) of the variable x is then known as a probability density function.

The definition is:

$$E\,(g(x)) = \int_{x_{min}}^{x_{max}} g(x) \cdot f(x)\, dx \qquad (2.5.1)$$

E(g(x)) is the mathematical expectation of the function g(x). If g(x) is a power function $g(x) = x^k$, then the expectation of g(x) is the k-th moment of the random variable x and is given by

$$E\,(x^k) = \int_{x_{min}}^{x_{max}} x^k \cdot f(x)\,dx \qquad (2.5.2)$$

k = 1 results in the important special case

$$E(x) = \int_{x_{min}}^{x_{max}} x \cdot f(x)\,dx \qquad (2.5.3)$$

E(x) is the first moment of the random variable x and is identical with the arithmetic mean μ of the distribution density function f(x) (equation 2.4.3)

$$E(x) = \mu \qquad (2.5.4)$$

In many cases $E(x) = \mu$ is called the true value of the quantity x, because it is assumed that due to errors of measurement the variable x fluctuates about μ.

It is a main task of the statistics of sampling to infer the true value from individual values of x.

The second moment is obtained analogously:

$$E\,(x^2) = \int_{x_{min}}^{x_{max}} x^2 \cdot f(x)\,dx \qquad (2.5.5)$$

$E(x^2)$ is the quadratic mean of the random variable x.

The expectation is also called the k-th central moment, if we set the function $g(x) = (x-\mu)^k$

$$E\,\{(x-\mu)^k\} = \int_{x_{min}}^{x_{max}} (x-\mu)^k \cdot f(x)\,dx \qquad (2.5.6)$$

The first central moment k = 1 results in:

$$E(x-\mu) = \int_{x_{min}}^{x_{max}} (x-\mu) \cdot f(x)\,dx = \int_{x_{min}}^{x_{max}} x\,f(x)\,dx - \mu \cdot \int_{x_{min}}^{x_{max}} f(x)\,dx$$

$$E\,(x-\mu) = 0 \tag{2.5.7}$$

The first central moment is always equal to zero.

The second central moment k = 2 results in:

$$E\;\{(x-\mu)^2\} = \int_{x_{min}}^{x_{max}} (x-\mu)^2 \cdot f(x)\,dx \tag{2.5.8}$$

The second central moment is identical with the variance σ^2 from equation (2.4.5). Just as was the case there, it is the measure of the width of the probability density function f(x)

$$E\;(\,(x-\mu)^2\,) = \sigma^2(x) \tag{2.5.9}$$

From the definition (2.5.8) the following rule for computation can be derived:

$$E(\,(x-\mu)^2\,) = \int_{x_{min}}^{x_{max}} (x^2 - 2\,\mu\,x + \mu^2) \cdot f(x)\,dx = \int_{x_{min}}^{x_{max}} x^2 \cdot f(x)\,dx - 2\,\mu \cdot \int_{x_{min}}^{x_{max}} x \cdot f(x)\,dx + \mu^2$$

and hence:

$$\sigma^2(x) = E(x^2) - E^2(x) \tag{2.5.10}$$

This rule of computation is extraordinarily important for the following derivations. If $E(x^2)$ and E(x) can be calculated, then $\sigma^2(x)$ can be determined from them. If, alternatively $\sigma^2(x)$ is known, then the expectation $E(x^2)$ can be derived by transformation:

$$E(x^2) = E^2(x) + \sigma^2(x) \tag{2.5.11}$$

Since $\sigma^2(x)$ is always a positive quantity, the quadratic mean $E(x^2)$ is always larger than the square of the arithmetic mean $E(x)$.

1) Rectangular distribution (uniform distribution), (Table 2.5.1, row 1)

$$f(x) = \begin{cases} 0 & x < -h/2 \\ 1/h & -h/2 \le x \le h/2 \\ 0 & x > h/2 \end{cases}$$

Then:

$$\int_{-h/2}^{h/2} f(x)\,dx = \frac{1}{h} \int_{-h/2}^{h/2} dx = 1$$

The distribution is normalized.

The expectation $E(x)$ is given by:

$$E(x) = \int_{-h/2}^{h/2} x \cdot f(x)\,dx = \frac{1}{h}\left(\frac{h^2}{8} - \frac{h^2}{8}\right) = 0$$

Worked examples for various distribution functions f(x):

Table 2.5.1 Expectations and variance of various distribution functions

Distribution function	E(x)	$E(x^2)$	Variance
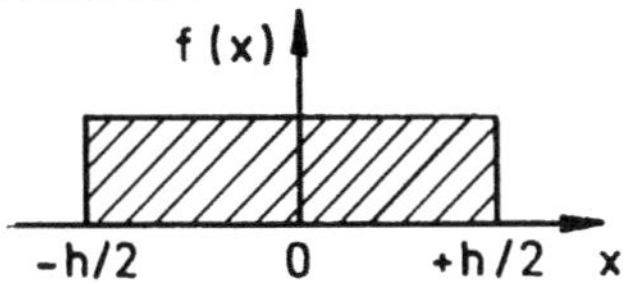	$E(x) = 0$ symmetrical about the origin	$E(x^2) = \frac{h^2}{12}$	$\sigma^2(x) = \frac{h^2}{12}$
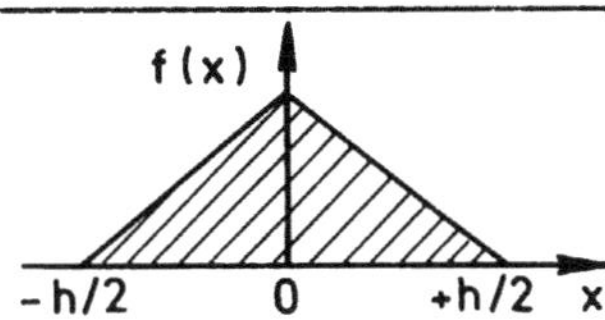	$E(x) = 0$ symmetrical about the origin	$E(x^2) = \frac{h^2}{24}$	$\sigma^2(x) = \frac{h^2}{24}$
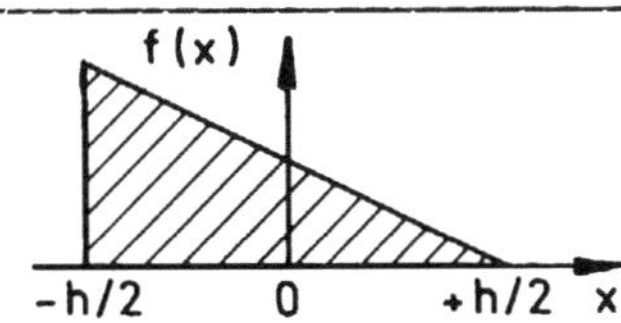	$E(x) = -\frac{h}{6}$	$E(x^2) = \frac{h^2}{12}$	$\sigma^2(x) = \frac{h^2}{18}$
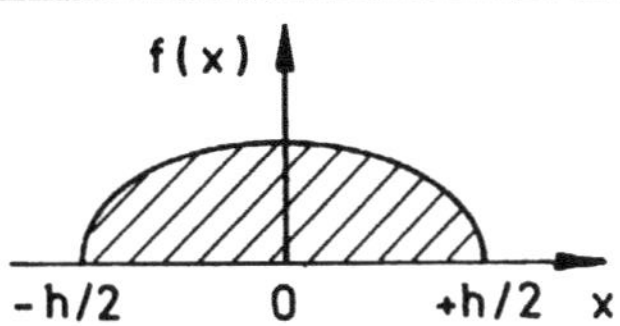	$E(x) = 0$ symmetrical about the origin	$E(x^2) = \frac{h^2}{16}$	$\sigma^2(x) = \frac{h^2}{16}$

The expectation $E(x^2)$ is given by:

$$E(x^2) = \int_{-h/2}^{h/2} x^2 \cdot f(x)\,dx = \frac{1}{h}\left(\frac{h^3}{24} + \frac{h^3}{24}\right) = \frac{h^2}{12}$$

The variance:

$$\sigma^2(x) = E(x^2) - E^2(x) = \frac{h^2}{12}$$

2) Triangular distribution (Table 2.5.1, row 2)

$$f(x) = \begin{cases} 0 & x < -h/2 \\ +\dfrac{4}{h^2}x + \dfrac{2}{h} & -\dfrac{h}{2} \leq x \leq 0 \\ -\dfrac{4}{h^2}x + \dfrac{2}{h} & 0 \leq x \leq \dfrac{h}{2} \\ 0 & x > \dfrac{h}{2} \end{cases}$$

Then:

$$\int_{-h/2}^{h/2} f(x)\,dx = \int_{-h/2}^{0} \left(\frac{4}{h^2}x + \frac{2}{h}\right) dx + \int_{0}^{h/2} \left(-\frac{4}{h^2} + \frac{2}{h}\right) dx$$

$$= \frac{2}{h^2}x^2\Bigg|_{-h/2}^{0} + \frac{2}{h}x\Bigg|_{-h/2}^{0} - \frac{2}{h^2}x^2\Bigg|_{0}^{h/2} + \frac{2}{h}x\Bigg|_{0}^{h/2}$$

$$= -\frac{1}{2} + 1 - \frac{1}{2} + 1 = 1$$

The distribution is normalized.

The expectation E(x) is given by:

$$E(x) = \int_{-h/2}^{0} \left(\frac{4}{h^2} x + \frac{2}{h}\right) x \, dx + \int_{0}^{h/2} \left(-\frac{4}{h^2} x + \frac{2}{h}\right) x \, dx$$

$$= \frac{4}{3h^2} x^3 \Big|_{-h/2}^{0} + \frac{1}{h} x^2 \Big|_{-h/2}^{0} - \frac{4}{3h^2} x^3 \Big|_{0}^{h/2} + \frac{1}{h} x^2 \Big|_{0}^{h/2}$$

$$= +\frac{h}{6} - \frac{h}{4} - \frac{h}{6} + \frac{h}{4} = 0$$

The expectation $E(x^2)$ is given by:

$$E(x^2) = \int_{-h/2}^{0} \left(\frac{4}{h^2} x + \frac{2}{h}\right) x^2 \, dx + \int_{-h/2}^{0} \left(-\frac{4}{h^2} x + \frac{2}{h}\right) x^2 \, dx$$

$$= \frac{1}{h^2} x^4 \Big|_{-h/2}^{0} + \frac{2}{3h} x^3 \Big|_{-h/2}^{0} - \frac{1}{h^2} x^4 \Big|_{0}^{h/2} + \frac{2}{3h} x^3 \Big|_{0}^{h/2}$$

$$= -\frac{h^2}{16} + \frac{h^2}{12} - \frac{h^2}{16} + \frac{h^2}{12} = \frac{h^2}{24}$$

The variance:

$$\sigma^2(x) = E(x^2) - E^2(x) = \frac{h^2}{24}$$

3) Triangular distribution (Table 2.5.1, row 3)

$$f(x) = \begin{cases} 0 & x < -\frac{h}{2} \\ -\frac{2}{h^2} x + \frac{1}{h} & -\frac{h}{2} \le x \le \frac{h}{2} \\ 0 & x > \frac{h}{2} \end{cases}$$

Then:

$$\int_{-h/2}^{h/2} f(x)\,dx = \int_{-h/2}^{h/2} \left(-\frac{2}{h^2}x + \frac{1}{h}\right) dx$$

$$= -\frac{1}{h^2}x^2\Big|_{-h/2}^{h/2} + \frac{1}{h}x\Big|_{-h/2}^{h/2}$$

$$= -\frac{1}{4} + \frac{1}{4} + \frac{1}{2} + \frac{1}{2} = 1$$

The distribution is normalized.

The expectation E(x) is given by:

$$E(x) = \int_{-h/2}^{h/2} \left(-\frac{2}{h^2}x + \frac{1}{h}\right) x\,dx$$

$$= -\frac{2}{2\,h^2}x^3\Big|_{-h/2}^{h/2} + \frac{1}{2\,h}x^2\Big|_{-h/2}^{h/2}$$

$$= -\frac{h}{12} - \frac{h}{12} + \frac{h}{8} - \frac{h}{8} = -\frac{h}{6}$$

The expectation $E(x^2)$ is given by:

$$E(x^2) = \int_{-h/2}^{h/2} \left(-\frac{2}{h^2}x + \frac{1}{h}\right) x^2\,dx$$

$$= -\frac{1}{2\,h^2}x^4\Big|_{-h/2}^{h/2} + \frac{1}{3\,h}x^3\Big|_{-h/2}^{h/2}$$

$$= -\frac{h^2}{16} + \frac{h^2}{16} + \frac{h^2}{24} + \frac{h^2}{24} = \frac{h^2}{12}$$

The variance:

$$\sigma^2(x) = E(x^2) - E^2(x)$$

$$\sigma^2(x) = \frac{h^2}{12} - \frac{h^2}{36} = \frac{h^2}{18}$$

4) Ellipse (Table 2.5.1, row 4)

$$f(x) = \begin{cases} 0 & x < -\frac{h}{2} \\ \frac{4}{h\cdot\pi}\sqrt{1-\left(\frac{2x}{h}\right)^2} & -\frac{h}{2} \leq x \leq \frac{h}{2} \\ 0 & x > \frac{h}{2} \end{cases}$$

Then:

$$\int_{-h/2}^{h/2} f(x)\,dx = \frac{4}{h\cdot\pi}\int_{-h/2}^{h/2}\sqrt{1-\left(\frac{2x}{h}\right)^2}\,dx = \frac{2}{\pi}\int_{-\pi/2}^{\pi/2}\cos^2\varphi\;d\varphi$$

$$= \frac{2}{\pi}\left(\frac{1}{2}\sin\varphi\cdot\cos\varphi + \frac{1}{2}\varphi\right)\Bigg|_{-\pi/2}^{\pi/2} = 1$$

The distribution is normalized.

The expectation E(x) is given by:

$$E(x) = \int_{-h/2}^{h/2}\frac{4}{h\cdot\pi}\sqrt{1-\left(\frac{2x}{h}\right)^2}\cdot x\,dx = \frac{h}{\pi}\cdot\int_{-\pi/2}^{\pi/2}\cos^2\varphi\cdot\sin\varphi\;d\varphi$$

$$= -\frac{h}{\pi}\cdot\frac{1}{3}\cos^3\varphi\;\Bigg|_{-\pi/2}^{\pi/2} = 0$$

The expectation $E(x^2)$ is given by:

$$E(x^2) = \frac{4}{h\cdot\pi} \cdot \int_{-h/2}^{h/2} \sqrt{1 - (\frac{2x}{h})^2} \cdot x^2 \, dx = \frac{h^2}{2\pi} \cdot \int_{-\pi/2}^{\pi/2} \cos^2\varphi \cdot \sin^2\varphi \, d\varphi$$

$$= -\frac{h^2}{16\pi} \left\{ \frac{1}{4} \sin 4\varphi - \varphi \right\} \Bigg|_{-\pi/2}^{\pi/2}$$

$$= -\frac{h^2}{16\pi} \{ -\pi/2 - \pi/2 \} = \frac{h^2}{16}$$

The variance:

$$\sigma^2(x) = E(x^2) - E^2(x) = \frac{h^2}{16}$$

2.6 Two-dimensional distribution functions

The elements of a population have up to now only been classified according to one attribute x. In the following text, the elements will be differentiated according to two attributes x and y. If both these quantities are random quantities, then each will have an arithmetic mean, μ for the attribute x and η for the attribute y, and correspondingly two variances, $\sigma^2(x)$ and $\sigma^2(y)$. An element is described unambiguously by the number pair x_i, y_i. In the case of comminuted ore, for example, the individual elements (particles) have on the one hand a certain particle size x_i, and on the other a certain metal content y_j. Each element for which the attribute x and y have the realizations x_i and y_j occurs at a certain relative frequency

$$h(x_i, y_j) = \frac{\text{number of elements with } x_i, y_j}{\text{total number of elements}} \qquad (2.6.1)$$

If the realizations x_i and y_j can be arranged in order according to their size, then a spatial bar chart is chosen for the graphical representation (Figure 2.6.1).

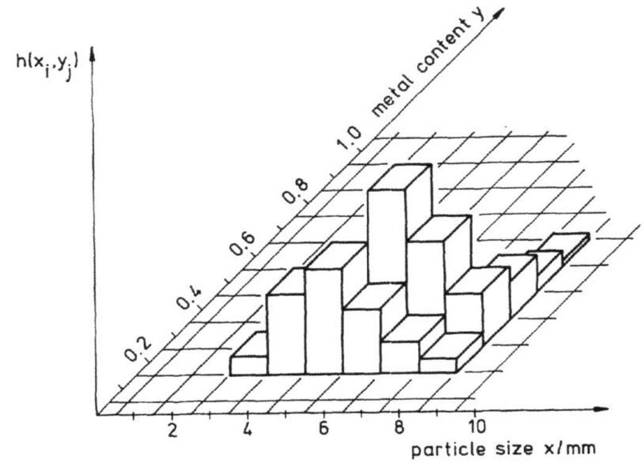

Figure 2.6.1 Bar chart of a two-dimensional frequency distribution

There is a two-dimensional discrete cumulative frequency distribution $H(x_i, y_j)$. $H(x_i, y_j)$ is the set of all elements, the realizations of which are $x \leqq x_i$ as well as $y \leqq y_j$. With the differential gradation of the attribute the discrète cumulative frequency distribution $H(x_i, y_i)$ becomes the two-dimensional cumulative distribution curve $F(x,y)$.

If $F(x, y)$ is sufficiently smooth, then a two-dimensional density distribution is given analogously to the one-dimensional density distributions by partial differentiation with respect to x and y:

$$f(x,y) = \frac{\partial^2 F(x, y)}{\partial x \partial y} \qquad (2.6.2)$$

The converse holds:

$$F(x, y) = \int_{x_{min}}^{x} \int_{y_{min}}^{y} f(\xi, \zeta)\, d\xi \, d\zeta \qquad (2.6.3)$$

2.7 Expectations and variances of a two-dimensional distribution function

If x and y are random variables, then the function g(x, y) is also a random variable. The distribution function f(x, y) of the variables x and y is then called a two-dimensional probability density function.

It is possible to calcultate:

$$E\,(g(x,y)\,) = \int_{x_{min}}^{x_{max}} \int_{y_{min}}^{y_{max}} g(x,\, y) \cdot f(x,\, y)\, dx\, dy \qquad (2.7.1)$$

E(g(x,y)) is the mathematical expectation of the two-dimensional function g(x, y).

If g(x, y) is a power function $g(x, y) = x^e.y^k$, then the expectation of g(x, y) is called the moments of the variables (x, y).

There are two first moments:

$$E(x) = \int_{x_{min}}^{x_{max}} \int_{y_{min}}^{y_{max}} x \cdot f(x,\, y)\, dx\, dy \qquad (2.7.2)$$

$$E(y) = \int_{x_{min}}^{x_{max}} \int_{y_{min}}^{y_{max}} y \cdot f(x,\, y)\, dx\, dy \qquad (2.7.3)$$

The second moments are given analogously:

$$E(x^2) = \int_{x_{min}}^{x_{max}} \int_{y_{min}}^{y_{max}} x^2 \cdot f(x,y)\, dy\, dx \qquad (2.7.4)$$

$$E(y^2) = \int_{y_{min}}^{y_{max}} \int_{x_{min}}^{x_{max}} y^2 \cdot f(x,y)\, dy\, dx \qquad (2.7.5)$$

$$E(x.y) = \int_{x_{min}}^{x_{max}} \int_{y_{min}}^{y_{max}} x \cdot y \cdot f(x, y)\, dy\, dx \qquad (2.7.6)$$

$E(x^2)$ and $E(y^2)$ correspond to the quadratic mean of the variables x and y in the population, $E(x.y)$ corresponding to the mean product of the two quantities.

Analogously to the one-dimensional distributions, central moments are defined by $g(x, y) = (x-E(x))^e.(y-E(y))^k$; there are therefore three second central moments:

$$\sigma^2(x) = \int_{x_{min}}^{x_{max}} \int_{y_{min}}^{y_{max}} (x - E(x))^2 \cdot f(x,y)\, dy\, dx = E(x^2) - E^2(x) \qquad (2.7.7)$$

$$\sigma^2(y) = \int_{x_{min}}^{x_{max}} \int_{y_{min}}^{y_{max}} (y - E(y))^2 \cdot f(x,y)\, dy\, dx = E(y^2) - E^2(y) \qquad (2.7.8)$$

$$Cov(x,y) = \int_{x_{min}}^{x_{max}} \int_{y_{min}}^{y_{max}} (y-E(y)) \cdot (x-E(x)) \cdot f(x,y)\, dy\, dx = E(x \cdot y)-E(x) \cdot E(y)$$

(2.7.9)

$Cov(x,y)$ is called the covariance of x and y, it being a measure of the dependence of random variables. For independent quantities x and y $Cov(x,y) = 0$. If the quantities x and y are equal, then $Cov(x,y) = \sigma^2(x) = \sigma^2(y)$.

The quantity is defined by:

$$\rho = \frac{\mathrm{Cov}\,(x,y)}{\sigma(x)\,\sigma(y)} \qquad \begin{array}{l} \sigma(x) > 0 \\ \sigma(y) > 0 \end{array} \qquad (2.7.10)$$

ρ is called the correlation coefficient. For independent quantities it is exactly equal to zero, just like Cov(x,y). The converse of these statements is not true. If Cov(x,y) or $\rho = 0$, then quantities x and y are called uncorrelated. All independent quantities are uncorrelated. The following expression holds:

$$-1 \leq \rho \leq +1 \qquad (2.7.11)$$

Limits ± 1 hold for a linear dependence of the random quantities x and y.

Worked examples for three distribution functions f(x,y)

1) Cuboid distribution (uniform distribution) (Table 2.7.1, row 1)

$$f(x,y) = \begin{cases} 0 & \text{if } x < -\frac{h_1}{2} \text{ or } y < -\frac{h_2}{2} \text{ or both} \\ \frac{1}{h_1\,h_2} & \text{if } -\frac{h_1}{2} \leq x \leq \frac{h_1}{2} \text{ and } -\frac{h_2}{2} \leq y \leq \frac{h_2}{2} \\ 0 & \text{if } \frac{h_1}{2} < x \text{ or } \frac{h_2}{2} < y \text{ or both} \end{cases}$$

<u>Table 2.7.1</u> Expectations and variances of two-dimensional distribution functions

	E(x)	E(y)	$\sigma^2(x)$	$\sigma^2(y)$	Cov(x, y)	E(x|y)	$\sigma^2(x\|y)$
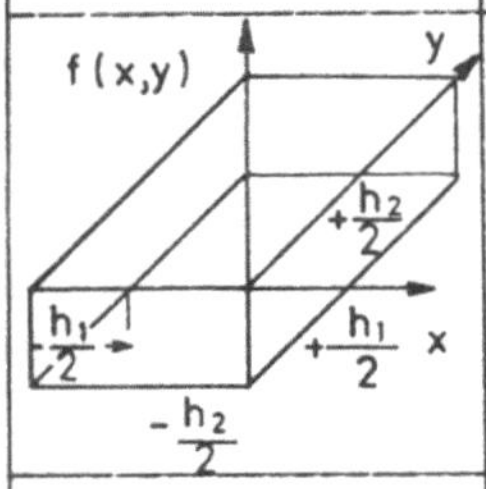	0	0	$\frac{h_1^2}{12}$	$\frac{h_2^2}{12}$	0	0	$\frac{h_1^2}{12}$
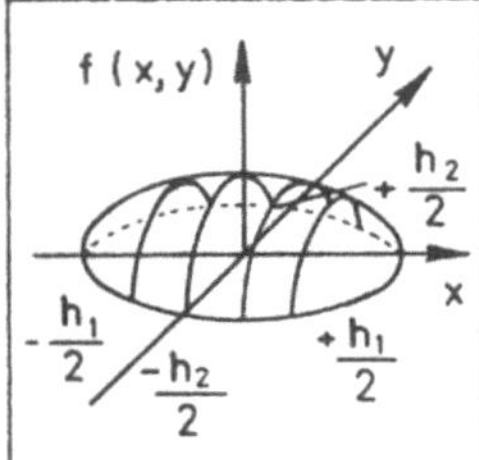	0	0	$\frac{h_1^2}{20}$	$\frac{h_2^2}{20}$	0	0	$\frac{h_1^2}{16} \cdot (1-(\frac{2y}{h_2})^2)$
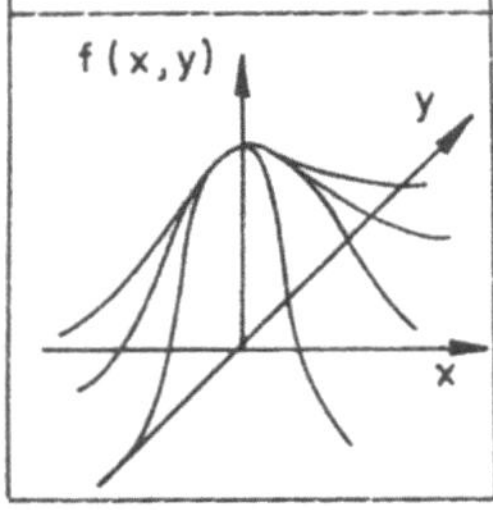	0	0	$\sigma^2 = a^2$	$\sigma^2 = a^2$	$a^2 \cdot b$ $b = \rho$	$\rho \cdot y$	$\sigma^2(1-\rho^2)$

Then:

$$\int_{-h_1/2}^{h_1/2} \int_{-h_2/2}^{h_2/2} \frac{1}{h_1 \cdot h_2} \cdot dx\, dy = 1$$

The distribution is normalized.

The expectation E(x) is given by:

$$E(x) = \int_{-h_1/2}^{h_1/2} \int_{-h_2/2}^{h_2/2} \frac{1}{h_1 \cdot h_2} \cdot x \, dy \, dx = \frac{1}{h_1} \int_{-h_1/2}^{h_1/2} x \, dx = 0$$

The expectation E(y) is given by:

$$E(y) = \int_{-h_1/2}^{h_1/2} \int_{-h_2/2}^{h_2/2} \frac{1}{h_1 \, h_2} \cdot y \, dy \, dx = \frac{1}{h_1 \, h_2} \int_{-h_1/2}^{h_1/2} \int_{-h_2/2}^{h_2/2} y \, dy \, dx = 0$$

The expectation $E(x^2)$ is given by:

$$E(x^2) = \int_{-h_1/2}^{h_1/2} \int_{-h_2/2}^{h_2/2} \frac{1}{h_1 \, h_2} \, x^2 \, dy \, dx = \frac{1}{h_1} \cdot \int_{-h_1/2}^{h_1/2} x^2 \, dx = \frac{h_1^2}{12}$$

The expectation $E(y^2)$ is given by:

$$E(y^2) = \int_{-h_1/2}^{h_1/2} \int_{-h_2/2}^{h_2/2} \frac{1}{h_1 \, h_2} \, y^2 \, dy \, dx = \frac{h_2^2}{12} \, \frac{1}{h_1} \cdot \int_{-h_1/2}^{h_1/2} dx = \frac{h_2^2}{12}$$

The expectation E(x.y) is given by:

$$E(x \cdot y) = \int_{-h_1/2}^{h_1/2} \int_{-h_2/2}^{h_2/2} \frac{1}{h_1 \, h_2} \, x \cdot y \, dy \, dx = \frac{1}{h_1 \, h_2} \int_{-h_1/2}^{h_1/2} \times \int_{-h_2/2}^{h_2/2} y \, dy \, dx = 0$$

The variance $\sigma^2(x)$ is given by:

$$\sigma^2(x) = E(x^2) - E^2(x) = \frac{h_1^2}{12}$$

The variance $\sigma^2(y)$ is given by:

$$\sigma^2(y) = E(y^2) - E^2(y) = \frac{h_2^2}{12}$$

The covariance:

$$Cov(x,y) = E(x \cdot y) - E(x) \cdot E(y) = 0$$

2) Ellipsoid distribution (Table 2.7.1, row 2)

$$f(x,y) = \frac{6}{\pi h_1 h_2} \sqrt{1 - (\frac{2x}{h_1})^2 - (\frac{2y}{h_2})^2}$$

$$\text{for } \left\{ 1 - (\frac{2x}{h_1})^2 - (\frac{2y}{h_2})^2 \right\} \geq 0$$

$f(x,y) = 0$ otherwise

Then:

$$\int_{-h_1/2}^{h_1/2} \int_{-h_2/2}^{h_2/2} \frac{6}{\pi h_1 h_2} \sqrt{1 - (\frac{2x}{h_1})^2 - (\frac{2y}{h_2})^2}\, dy\, dx =$$

$$= \frac{3}{\pi h_2} \cdot \int_{-\frac{h_2}{2}}^{h_2/2} (1 - (\frac{2y}{h_2})^2) \int_{-\pi/2}^{\pi/2} \cos^2\varphi \, d\varphi \quad dy =$$

$$= \frac{3}{4} \cdot \int_{-\pi/2}^{\pi/2} \cos^3\varphi \, d\varphi = \frac{3}{4} \left| \sin\varphi - \frac{1}{3}\sin^3\varphi \right|_{-\pi/2}^{\pi/2} = 1$$

The distribution is normalized.

Expectations E(x) and E(y) are given by:

$$E(x) = \int_{-h_1/2}^{h_1/2} \int_{-h_2/2}^{h_2/2} \frac{6}{\pi h_1 h_2} \sqrt{1 - (\frac{2x}{h_1})^2 - (\frac{2y}{h_2})^2} \cdot x \, dy \, dx$$

$$= \frac{3}{\pi h_2} \int_{-h_2/2}^{h_2/2} (1 - (\frac{2y}{h_2})^2)^{3/2} \cdot \frac{h_1}{2} \cdot \int_{-\pi/2}^{\pi/2} \cos^2\varphi \cdot \sin\varphi \; d\varphi \; dy = 0$$

Analogously:

$$E(y) = 0$$

Expectations $E(x^2)$ and $E(y^2)$ are given by:

$$E(x^2) = \int_{-h_1/2}^{h_1/2} \int_{-h_2/2}^{h_2/2} \frac{6}{\pi h_1 h_2} \sqrt{1 - (\frac{2x}{h_1})^2 - (\frac{2y}{h_2})^2} \cdot x^2 \, dy \, dx =$$

$$= \frac{3}{\pi h_2} \cdot \int_{-h_2/2}^{h_2/2} (1 - (\frac{2y}{h_2})^2)^2 \cdot (\frac{h_1}{2})^2 \cdot \int_{-\pi/2}^{\pi/2} \cos^2\varphi \sin^2\varphi \; d\varphi \, dy =$$

$$= \frac{h_1^2}{20}$$

and analogously

$$E(y^2) = \frac{h_2^2}{20}$$

Expectation E(x.y) is given by:

$$E(x\cdot y) = \frac{6}{\pi h_1 h_2} \int_{-h_1/2}^{h_1/2} \int_{-h_2/2}^{h_2/2} \sqrt{1 - (\frac{2x}{h_1})^2 - (\frac{2y}{h_2})^2} \cdot x\cdot y \, dy \, dx$$

$$= \frac{6}{\pi h_1 h_2} \int_{-h_2/2}^{h_2/2} y \int_{-h_1/2}^{h_1/2} \sqrt{1 - (\frac{2x}{h_1})^2 - (\frac{2y}{h_2})^2} \; x \, dx \, dy$$

$$= \frac{3}{\pi h_2} \int_{-h_2/2}^{h_2/2} y \, (1 - (\frac{2y}{h_2})^2)^{\frac{3}{2}} \cdot \frac{h_1}{2} \cdot \int_{-\pi/2}^{\pi/2} \cos^2\varphi \cdot \sin\varphi \, d\varphi \, dy = 0$$

The variances $\sigma^2(x)$, $\sigma^2(y)$ are given by:

$$\sigma^2(x) = E(x^2) - E^2(x) = \frac{h_1^2}{20}$$

$$\sigma^2(y) = E(y^2) - E^2(y) = \frac{h_2^2}{20}$$

The covariance:

$$\text{Cov}(x, y) = E(x,y) - E(x) \cdot E(y) = 0$$

3) Two-dimensional normal distribution (Table 2.7.1, row 3)

$$f(x,y) = \frac{1}{2\pi a^2 \sqrt{1-b^2}} \cdot e^{-\frac{h(x,y)}{2}} \qquad \begin{matrix} -\infty \le x \le \infty \\ -\infty \le y \le \infty \end{matrix}$$

$$\text{with } h(x,y) = \frac{1}{1-b^2} \left\{ (\frac{x}{a})^2 - 2b\frac{x\cdot y}{a^2} + (\frac{y}{a})^2 \right\}$$

$$= \frac{1}{1-b^2} \left\{ (\frac{x}{a})^2 - 2b\frac{x\cdot y}{a^2} + (\frac{b\cdot y}{a})^2 - (\frac{b\,y}{a})^2 + \frac{y^2}{a^2} \right\}$$

$$= \frac{1}{1-b^2} \left\{ \frac{(x-b\,y)^2}{a^2} + \frac{y^2}{a^2}(1-b^2) \right\}$$

Then:

$$\int_{-\infty}^{+\infty}\int_{-\infty}^{+\infty} \frac{1}{2\pi a^2\sqrt{1-b^2}} \cdot e^{-\frac{h(x,y)}{2}}\, dx\, dy =$$

$$\frac{1}{2\pi a^2\sqrt{1-b^2}} \int_{-\infty}^{+\infty} e^{-\frac{1}{2}\left(\frac{y}{a}\right)^2} \int_{-\infty}^{+\infty} e^{-\frac{1}{2}\left(\frac{x-by}{a\sqrt{1-b^2}}\right)^2} dx\, dy =$$

$$\frac{1}{\sqrt{2\pi}\cdot a} \cdot \int_{-\infty}^{+\infty} e^{-\frac{1}{2}\left(\frac{y}{a}\right)^2} dy = 1$$

The distribution is normalized.

The expectations E(x) and E(y) are given by:

$$E(y) = \frac{1}{2\pi a^2\sqrt{1-b^2}} \cdot \int_{-\infty}^{+\infty} y\, e^{-\frac{1}{2}\left(\frac{y}{a}\right)^2} \int_{-\infty}^{+\infty} e^{-\frac{1}{2}\left(\frac{x-by}{a\sqrt{1-b^2}}\right)^2} dx\, dy$$

$$E(y) = \frac{1}{\sqrt{2\pi}\, a} \int_{-\infty}^{+\infty} y \cdot e^{-\frac{1}{2}\left(\frac{y}{a}\right)^2} dy = 0$$

and analogously:

$E(x) = 0$

The expectations $E(x^2)$ and $E(y^2)$ are given by:

$$E(y^2) = \frac{1}{2\pi a^2\sqrt{1-b^2}} \int_{-\infty}^{+\infty} y^2 \cdot e^{-\frac{1}{2}\left(\frac{y}{a}\right)^2} \int_{-\infty}^{+\infty} e^{-\frac{1}{2}\left(\frac{x-by}{a\sqrt{1-b^2}}\right)^2} dx\, dy$$

$$E(y^2) = \frac{1}{\sqrt{2\pi}\cdot a} \int_{-\infty}^{+\infty} y^2 \cdot e^{-\frac{1}{2}\left(\frac{y}{a}\right)^2} dy = a^2$$

and analogously:

$$E(x^2) = a^2$$

The expectation E(x.y) is given by:

$$E(x\cdot y) = \frac{1}{2\,\pi\, a^2 \sqrt{1-b^2}} \int_{-\infty}^{+\infty} y \cdot e^{-\frac{1}{2}\left(\frac{y}{a}\right)^2} \int_{-\infty}^{+\infty} x \cdot e^{-\frac{1}{2}\left(\frac{x-by}{a\sqrt{1-b^2}}\right)^2} dx\, dy$$

$$E(x\cdot y) = \frac{b}{\sqrt{2\,\pi}\cdot a} \int_{-\infty}^{+\infty} y^2 \cdot e^{-\frac{1}{2}\left(\frac{y}{a}\right)^2} dy = a^2 \cdot b$$

The variances:

$$\sigma^2(x) = E(x^2) - E^2(x) = a^2$$

$$\sigma^2(y) = E(y^2) - E^2(y) = a^2$$

The covariance:

$$\mathrm{Cov}\,(x,\, y) = E(x\cdot y) - E(x) \cdot E(y) = a^2 \cdot b$$

and therefore for the correlation coefficient:

$$\rho = \frac{\mathrm{Cov}(x\cdot y)}{\sigma(x)\cdot\sigma(y)} = b$$

2.8 Conditional expectations

The two-dimensional distribution density function f(x,y) covers the plane of the variables (x,y) (Figure 2.8.1). If all the values with a constant variable y are singled out, then a section through the probability hill results.

If the resulting section function is normalized, then the expectation and the variance of this new distribution f(x y) can be determined. f(x|y) is called the conditional distribution density function.

$$f(x|y) = \frac{f(x,y)}{\int_{x_{min}}^{x_{max}} f(x,y)\,dx}$$

If g(x,y) is a function of the variables x and y, then

$$E(g\,(x,y)|y) = \int_{x_{min}}^{x_{max}} g(x,y) \cdot f(x|y)\,dx \qquad (2.8.1)$$

is the conditional expectation of the function g(x,y).

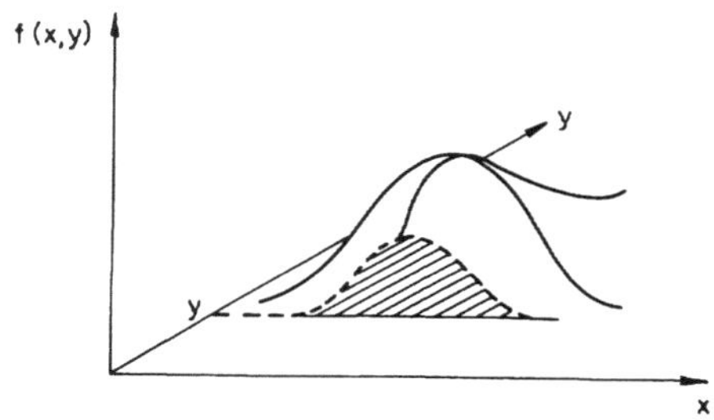

Figure 2.8.1 Diagram to explain conditional distributions

The "condition", expressed by the vertical stroke followed by y, implies that a new population has been formed from all elements for wich y has a certain realization. The conditional expectations are therefore still functions of the condition.

Therefore:

$$E(x\,|y) = \int_{x_{min}}^{x_{max}} x \cdot f(x\,|y)\,dx \qquad (2.8.2)$$

is the conditional expectation of the quantity x. For the conditional variance correspondingly:

$$\sigma^2(x|y) = \int_{x_{min}}^{x_{max}} (x - E(x|y))^2 \cdot f(x|y)dx \qquad (2.8.3)$$

$$\sigma^2(x|y) = E(x^2|y) - E^2(x|y) \qquad (2.8.4)$$

<u>Worked examples for conditional expectations and variances</u>

<u>1. Cuboid distribution (uniform distribution) (Table 2.7.1, row 1)</u>

$$f(x,y) = \frac{1}{h_1\ h_2} \text{ in the region } \begin{cases} -\frac{h_1}{2} \leq x \leq \frac{h_1}{2} \\ -\frac{h_2}{2} \leq y \leq \frac{h_2}{2} \end{cases}$$

Normalization of the conditional distribution function $f(x|y)$:

$$f(x|y) = \frac{f(x,y)}{\int_{x_{min}}^{x_{max}} f(x,y)\,dx} = \frac{\frac{1}{h_1\ h_2}}{\frac{1}{h_1\ h_2}\int_{-h_1/2}^{h_1/2} dx} = \frac{1}{h_1}$$

The conditional expectation $E(x|y)$ is given by:

$$E(x|y) = \int_{-h_1/2}^{h_1/2} x \cdot \frac{1}{h_1}\ dx = \frac{1}{h_1} \cdot \frac{x^2}{2}\Bigg|_{-h_1/2}^{h_1/2} = 0$$

The conditional expectation $E(x^2|y)$ is given by:

$$E(x^2|y) = \frac{1}{h_1} \cdot \int_{-h_1/2}^{h_1/2} x^2\ dx = \frac{1}{h_1}\ \frac{x^3}{3}\Bigg|_{-h_1/2}^{h_1/2} = \frac{h_1^{\ 2}}{12}$$

The conditional variance:

$$\sigma^2(x|y) = E(x^2|y) - E^2(x|y) = \frac{h_1^2}{12}$$

The conditional expectations and the conditional variances agree with the unconditional expectations, ie. the quantities are independent.

2. Ellipsoid distribution (Table 2.7.1, row 2)

$$f(x,y) = \frac{6}{\pi h_1 h_2} \sqrt{1 - (\frac{2x}{h_1})^2 - (\frac{2y}{h_2})^2} \text{ in the region } \begin{cases} -\frac{h_1}{2} \le x \le \frac{h_1}{2} \\ -\frac{h_2}{2} \le y \le \frac{h_2}{2} \end{cases}$$

Normalization of the conditional distribution function:

$$f(x|y) = \frac{f(x,y)}{\int_{x_{min}}^{x_{max}} f(x,y)\,dx} = \frac{\sqrt{1-(\frac{2x}{h_1})^2 - (\frac{2y}{h_2})^2}}{\frac{h_1}{2}\,(1-(\frac{2y}{h_2})^2) \int_{-\pi/2}^{+\pi/2} \cos^2\varphi \, d\varphi}$$

$$f(x|y) = \frac{4}{h_1 \pi} \frac{\sqrt{1-(\frac{2y}{h_2})^2 - (\frac{2x}{h_1})^2}}{(1-(\frac{2y}{h_2})^2)} .$$

The conditional expectation $E(x|y)$ is given by:

$$E(x|y) = \int_{x_{min}}^{x_{max}} x \cdot f(x|y)\,dx = \frac{2}{\pi}\sqrt{1-(\frac{2y}{h_2})^2} \cdot \int_{-\pi/2}^{\pi/2} \cos^2\varphi \sin\varphi \, d\varphi = 0$$

The conditional expectation $E(x^2|y)$ is given by:

$$E(x^2|y) = \int_{x_{min}}^{x_{max}} x^2 f(x|y)\, dx =$$

$$= \frac{2}{\pi} \left(\frac{h_1}{2}\right)^2 \left(1 - \left(\frac{2y}{h_2}\right)^2\right) \int_{-\pi/2}^{\pi/2} \cos^2\varphi \sin^2\varphi \; d\varphi$$

$$E(x^2|y) = \frac{h_1^2}{16} \left(1 - \left(\frac{2y}{h_2}\right)^2\right)$$

The conditional variance:

$$\sigma^2(x|y) = E(x^2|y) - E^2(x|y) = \frac{h_1^2}{16} \left(1 - \left(\frac{2y}{h_2}\right)^2\right)$$

<u>3. Two-dimensional normal distribution (Table 2.7.1, row 3)</u>

$$f(x,y) = \frac{1}{2\pi\sigma^2\sqrt{1-\rho^2}} \cdot e^{-\frac{h(x,y)}{2}} \qquad \begin{array}{l} -\infty < x < \infty \\ -\infty < y < \infty \end{array}$$

$$h(x,y) = \frac{1}{1-\rho^2} \left\{ \left(\frac{x}{\sigma}\right)^2 - 2\rho \frac{x \cdot y}{\sigma^2} + \left(\frac{y}{\sigma}\right)^2 \right\} = \frac{1}{1-\rho^2} \quad \frac{(x-\rho y)^2}{\sigma^2} + \frac{y^2}{\sigma^2} (1-\rho^2))$$

$$= \frac{1}{1-\rho^2} \left(\frac{x-\rho y}{\sigma}\right)^2 + \left(\frac{y}{\sigma}\right)^2$$

Normalization of the conditional distribution function:

$$f(x|y) = \frac{f(x,y)}{\int_{x_{min}}^{x_{max}} f(x,y)\, dx}$$

$$\int_{-\infty}^{+\infty} \frac{1}{2\pi\sigma^2\sqrt{1-\rho^2}} \cdot e^{-\frac{1}{2}\left(\frac{x-\rho y}{\sigma\sqrt{1-\rho^2}}\right)^2 - \frac{1}{2}\left(\frac{y}{\sigma}\right)^2} dx = \frac{1}{\sqrt{2\pi}\cdot\sigma} \cdot e^{-\frac{1}{2}\left(\frac{y}{\sigma}\right)^2}$$

$$f(x\,|\,y) = \frac{1}{\sqrt{2\pi}\cdot\sigma\sqrt{1-\rho^2}} \cdot e^{-\frac{h(x,y)}{2} + \frac{1}{2}\left(\frac{y}{\sigma}\right)^2}$$

The conditional expectation $E(x\,|\,y)$ is given by:

$$E(x\,|\,y) = \int_{-\infty}^{+\infty} x \cdot f(x\,|\,y)\, dx = \frac{e^{\frac{1}{2}\left(\frac{y}{\sigma}\right)^2}}{\sqrt{2\pi}\cdot\sigma\sqrt{1-\rho^2}} \int_{-\infty}^{+\infty} x \cdot e^{-\frac{h(x,y)}{2}} dx$$

$$= \frac{1}{\sqrt{2\pi}\cdot\sigma\sqrt{1-\rho^2}} \int_{-\infty}^{+\infty} x \cdot e^{-\frac{1}{2}\left(\frac{x-\rho y}{\sigma\sqrt{1-\rho^2}}\right)^2} dx$$

$$= \frac{1}{\sqrt{2\pi}} \int_{-\infty}^{+\infty} \left(\sigma\sqrt{1-\rho^2}\right)\cdot\xi + \rho y) \cdot e^{-\frac{1}{2}\xi^2} d\xi$$

$$E(x\,|\,y) = \rho \cdot y$$

The conditional expectation $E(x^2\,|\,y)$ is given by:

$$E(x^2\,|\,y) = \int_{-\infty}^{+\infty} x^2 f(x\,|\,y)\, dx = \frac{1}{\sqrt{2\pi}\,\sigma\sqrt{1-\rho^2}} \int_{-\infty}^{+\infty} x^2 \cdot e^{-\frac{1}{2}\left(\frac{x-\rho y}{\sigma\sqrt{1-\rho^2}}\right)} dx$$

$$= \frac{1}{\sqrt{2\pi}} \left\{ \int_{-\infty}^{+\infty} \sigma^2(1-\rho^2)\cdot\xi^2\; e^{-\frac{1}{2}\xi^2} d\xi + \rho^2 y^2 \int_{-\infty}^{+\infty} e^{-\frac{1}{2}\xi^2} d\xi \right\}$$

$$E(x^2\,|\,y) = \sigma^2(1-\rho^2) + \rho^2 \cdot y^2$$

The conditional variance:

$$\sigma^2(x|y) = E(x^2|y) - E^2(x|y) = \sigma^2(1-\rho^2) + \rho^2 \cdot y^2 - \rho^2 \cdot y^2$$

$$\sigma^2(x|y) = \sigma^2(1-\rho^2)$$

2.9 Rules for calculating expectations and variances

2.9.1 One-dimensional distribution functions (Table (2.9.1)

Rules for calculation can be derived from the difinitions of expectations and variances, with which the frequently long-winded derivation of the distribution function can be avoided.

If a and b are constant parameters, then:

$$E(a \pm b\,x) = a \pm b\,E(x) \tag{2.9.1}$$

$$E(x^2) = E^2(x)\left\{1 + \left(\frac{\sigma(x)}{E(x)}\right)^2\right\} \tag{2.9.2}$$

If the function g(x) can be expanded in the Taylor series about the expectation $E(x) = \mu$, which can be terminated after the third term, then the following approximation equation holds:

$$E(g(x)) \approx g(\mu) + \frac{1}{2}\,g''(\mu) \cdot \sigma^2(x) \tag{2.9.3}$$

where $g''(\mu)$ is the second differential of the function g(x) with respect to the argument x at $x = \mu$.

Table 2.9.1 Approximate calculation of $E(g(x))$ and $\sigma^2(g(x))$

	Function $g(x)$	Expectation $E(g(x))$	Variance $\sigma^2(g(x))$
1	Power function $g(x) = a \cdot x^k$ a = const. k = real $(x > 0)$	$E(a \cdot x^k) = a \cdot E^k(x)\,(1 + \frac{1}{2}k(k-1)(\frac{\sigma(x)}{E(x)})^2)$	$\sigma^2(a\,x^k) \approx \sigma^2(x) \cdot a^2 \cdot k^2 \cdot E^{2(k-1)}(x)$
2	$g(x) = a \cdot x^2$ $(x > 0)$	$E(a\,x^2) = a \cdot E^2(x)\,(1 + (\frac{\sigma(x)}{E(x)})^2)$	$\sigma^2(a\,x^2) \approx \sigma^2(x) \cdot a^2 \cdot 4 \cdot E^2(x)$
3	$g(x) = a \cdot \sqrt{x}$ $(x > 0)$	$E(a\sqrt{x}) \approx a\sqrt{E(x)}\,(1 - \frac{1}{8}(\frac{\sigma(x)}{E(x)})^2)$	$\sigma^2(a\sqrt{x}) \approx \sigma^2(x) \cdot a^2 \cdot \frac{1}{4}\,E^{-1}(x)$
4	$g(x) = \frac{a}{x}$ $(x > 0)$	$E(\frac{a}{x}) \approx \frac{a}{E(x)}\,(1 + (\frac{\sigma(x)}{E(x)})^2)$	$\sigma^2(\frac{a}{x}) \approx \sigma^2(x) \cdot \frac{a^2}{E^4(x)}$
5	$g(x) = \ln(\frac{x}{a})$ $x > 0$	$E(\ln(\frac{x}{a})) \approx \ln(\frac{E(x)}{a}) - \frac{1}{2}(\frac{\sigma(x)}{E(x)})^2$	$\sigma^2(\ln(\frac{x}{a})) \approx \frac{\sigma^2(x)}{E^2(x)}$
6	$g(x) = e^{x/a}$	$E(e^{x/a}) \approx e^{\frac{E(x)}{a}}\,(1 + \frac{1}{2}(\frac{\sigma(x)}{a})^2)$	$\sigma^2(e^{x/a}) \approx \sigma^2(x) \cdot \frac{e^{\frac{E(x)}{a}}}{a^2}$

The result for the variances is:

$$\sigma^2 (a + b x) = b^2 \cdot \sigma^2 (x) \qquad (2.9.4)$$

If the distribution function is not too asymmetrical, then with the aid of the Taylor expansion of the function g(x) we obtain approximately

$$\sigma^2 (g(x)) \approx g'^2 (\mu) \cdot \sigma^2 (x) \qquad (2.9.5)$$

where $g'(\mu)$ is the first differential of the function g(x) with respect to the argument x at $x = \mu$.

2.9.2 Two-dimensional distribution function (Table 2.9.2)

If the elements of a population are distinguished by two attributes x and y, then:

$$E(x \pm y) = E(x) \pm E(y) \qquad (2.9.6)$$

The expectation of a sum of arbitrary random quantities is equal to the sum of the expectations.

$$E(x \cdot y) = E(x) \cdot E(y) + Cov(x, y) \qquad (2.9.7)$$

If the quantities x and y are independent, then the covariance disappears and:

$$E(x \cdot y) = E(x) \cdot E(y) \qquad (2.9.8)$$

The expectation of a product of independent random quantities is equal to the product of the expectations.

The result for the variances is:

$$\sigma^2(x \pm y) = \sigma^2(x) + \sigma^2(y) \pm 2\,\mathrm{Cov}\,(x,y) \qquad (2.9.9)$$

If the quantities x and y are independent, then the equation simplifies to:

$$\sigma^2(x \pm y) = \sigma^2(x) + \sigma^2(y) \qquad (2.9.10)$$

For independent random quantities the variance of the sum is equal to the sum of the variances. If the function g(x,y) can be expanded in the two-dimensional Taylor series about the expectations E(x) = μ and E(y) = η, then the approximation is obtained:

$$E(g(x,y)) \approx g(\mu,\eta) + \frac{1}{2!}\left(\frac{\partial^2 g(\mu,\eta)}{\partial x^2}\cdot\sigma_x^2 + \frac{\partial^2 g(\mu,\eta)}{\partial y^2}\cdot\sigma_y^2 + 2\,\frac{\partial^2 g(\mu,\eta)}{\partial x\,\partial y}\cdot \mathrm{Cov}\,(x,y)\right) \qquad (2.9.11a)$$

$$\sigma^2(g(x,y)) \approx \left(\frac{\partial g(\mu,\eta)}{\partial x}\right)^2 \sigma^2(x) + \left(\frac{\partial g(\mu,\eta)}{\partial y}\right)^2\cdot\sigma^2(y) + 2\left(\frac{\partial g(\mu,\eta)}{\partial x}\cdot\frac{\partial g(\mu,\eta)}{\partial y}\right)\mathrm{Cov}\,(x,y)$$

(2.9.11 b)

For independent quantities x and y the equation 2.9.11 becomes the Gaussian law of propagation of errors:

$$\sigma^2(g(x,y)) = \left(\frac{\partial g(\mu,\eta)}{\partial x}\right)^2\cdot\sigma^2(x) + \left(\frac{\partial g(\mu,\eta)}{\partial y}\right)^2\cdot\sigma^2(y) \qquad (2.9.12)$$

Table 2.9.2 Approximation for two-dimensional expectations and variances

	Function $g(x,y)$	Expectation $E(g(x,y))$	Variance $\sigma^2(g(x,y))$
1	$g(x,y) = x \pm y$	$E(x \pm y) = E(x) \pm E(y)$	$\sigma^2(x \pm y) = \sigma^2(x) + \sigma^2(y) \pm 2\,\mathrm{Cov}\,(x,y)$
2	$g(x,y) = x \cdot y$	$E(x \cdot y) = E(x)\,E(y) + \mathrm{Cov}(x,y)$	$\sigma^2(x \cdot y) \approx E^2(y) \cdot \sigma^2(x) + E^2(x) \cdot \sigma^2(y) +$ $2\,E(x) \cdot E(y) \cdot \mathrm{Cov}\,(x,y)$
3	$g(x,y) = \frac{x}{y}$	$E(\frac{x}{y}) = \frac{E(x)}{E(y)}\ (1 + \frac{\sigma^2(y)}{E^2(y)}\ -$ $- \frac{\mathrm{Cov}(x,y)}{E(x) \cdot E(y)}$	$\sigma^2(\frac{x}{y}) \approx \frac{E^2(x)}{E^2(y)} \cdot \left\{ (\frac{\sigma(x)}{E(x)})^2 + (\frac{\sigma(y)}{E(y)})^2 \right.$ $\left. - \frac{2\,\mathrm{Cov}\,(x,y)}{E(x) \cdot E(y)} \right\}$

2.9.3 Conditional distribution functions

The conditional distribution function is a one-dimensional function with a "condition" kept constant, and the rules for expectations and variances are valid which were derived for one- dimensional functions:

$$E\,((a + b\,x)|y) = a + b\,E(x|y) \qquad (2.9.13)$$

$$E\,((y + x)\,|\,y) = y + E\,(x|y) \qquad (2.9.14)$$

The conditional expectation is only valid for a certain y. In general, a different expectation will be obtained for each different y. The conditional expectation is also usually a function of its condition:

$$E\,(g\,(x,y)|\,y\,) = g^{*}(y) \qquad (2.9.15)$$

If the variable y is itself a random quantity, then the expectation of the quantity g(y) can be formed. The following chain rule according to Fisz[3] applies:

$$E(g(x)\,) = E_y\;\{\,E\,(g(x)|\,y)\,\} \qquad (2.9.16)$$

E(g(x)) is the expectation of the function g(x) over the whole two-dimensional distribution function. The index y signifies that the conditional expectation E(g(x)y) is now only a function of y and no longer a function of x.

An analogous derivation for the variances leads to:

$$\sigma^2(x) = E_y\,(\sigma^2(x|y)\,) + \sigma_y^2\,(E(x|y)\,) \qquad (2.9.17)$$

Example: Two-dimensional normal distribution (Table 2.7.1, row 3)

$$E(x|y) = \rho \cdot y\;; \quad \sigma^2(x|y) = \sigma^2(1 - \rho^2)$$

Using equation (2.9.16)

$$E(x) = E_y\ (E(x|y)) = E\ (\rho \cdot y) = \rho \cdot E(y) = 0$$

Using equation (2.9.17)

$$\sigma^2(x) = E_y\ (\sigma^2\ (1-\rho^2)) + \sigma^2_y\ (\rho \cdot y) = \sigma^2(1-\rho^2) + \rho^2 \cdot \sigma^2$$

$$\sigma^2(x) = \sigma^2$$

This agrees with the parameters calculated in section 2.7.

Note: In reality nearly all expectations are conditional expectations, because the populations considered are usually subsets of larger sets. If it is assumed, however, that the subset on which the considerations are based always remains the same, then the condition does not matter. The rules for calculating conditional expectations are the same in this case as the rules for calculating unconditional expectations. Only if it is intended to draw inferences from the subset about a larger population, then the condition must be noted, and the equations (2.9.16) and (2.9.17) will be used as appropriate.

3 Random Sampling Distributions

3.1 Concept of the random sample

As already mentioned in section 2.1, in statistics a finite subset which consists of elements of the population is called a random sample. The number of elements in the sample is called the sample size or the sample extent. The sample extent can in the limiting case be just one element as well as the whole set. If it is assumed that the random sample can be repeated infinitely often under the same conditions, then the infinite number of realizations of the sample attribute form a random sampling distribution function. According to Bernoulli´s rule of large numbers, the sampling distribution function corresponds to a probability function as the limit of the frequency distribution of samples. If the sample consists of, for example, a throw of a true die and the number of dots is the realization of the sample attribute, the frequency distribution of Figure 2.1.1 will become the probability distribution of Figure 3.1.1 after an infinite number of throws.

In addition to the infinitely repeatable sample, the condition of random selection of the sample must be satisfied. A random selection means that the elements of a random sample have a well defined probability of being allocated to the sample. The probability of selecting the realization of a certain number of dots in the case of a true die is exactly 1/6.

The requirement that each random sample must be the result of random selection is frequently not so easily satisfied. Very frequently the so-called urn scheme is used to aid visualization.

An urn is a container in which the elements, for example, white and black balls are "well" mixed. The probability of findig a particle of one kind, independently of its neighbors, must be equal to the numeric concentration in the whole mixture. These

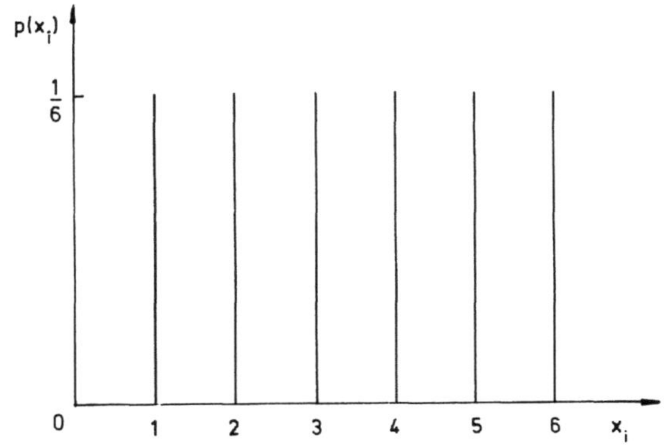

Figure 3.1.1 Probability distribution $p(x_i)$ of the realization x_i of a true die

elements are taken randomly from the urn. The sample therefore satisfies the condition of random selection. Two principal cases are distinguished.

A. Sampling with replacement:

After the realization of the attribute of the element has been noted, the element is replaced and the urn subjected to a good mixing. That is to say that the composition of the population is not changed by the sampling.

B. Sampling without replacement:

The elements which have been removed one by one by the process of sampling are not replaced. The composition of the population changes.

3.2 Sampling without replacement - hypergeometric distribution

If a sample of extent n is taken from an urn (population) containing N elements, of which N_x are white and N_y are black, without replacing the removed elements, then the probabilities $P_H(n_x)$ of obtaining exactly n_x elements form a hypgergeometric distribution. The classic definition of probabilities is used for the calculation.

$$P_4(n_x) = \frac{\text{Number of favorable outcomes}}{\text{Number of possible outcomes}} \qquad (3.2.1)$$

The number of all possible combinations of n elements, which can be removed from the population of N elements, is, regardless of sequence:

$$\frac{N!}{n!(N-n)!} = \binom{N}{n} \qquad (3.2.2)$$

The number of favorable outcomes still has to be determined. The number of possibilities of removing excatly n_x elements from the N_x white elements is:

$$\frac{N_x!}{n_x!\,(N_x-n_x)!} = \binom{N_x}{n_x} \qquad (3.2.3)$$

Similarly, the number of possibilities of removing n_y black elements from the N_y black elements is:

$$\frac{N_y!}{n_y!(N_y-n_y)!} = \frac{(N - N_x)!}{(n-n_y)!\,(N-N_x-n+n_x)} = \binom{N-N_x}{n-n_x} \qquad (3.2.4)$$

Therefore, the number of favorable outcomes, that is removing n_x white from N_x or n_y black from N_y, is:

$$\binom{N_x}{n_x} \cdot \binom{N_y}{n_y} = \binom{N_x}{n_x}\binom{N-N_x}{n-n_x} \qquad (3.2.5)$$

The probability distribution sought is found to be:

$$\boxed{P_H(n_x) = \frac{\binom{N_x}{n_x}\binom{N-N_x}{n-n_x}}{\binom{N}{n}}} \qquad (3.2.6)$$

This is the hypergeometric distribution. It is a discrete one-dimensional probability distribution (Figure 3.2.1).

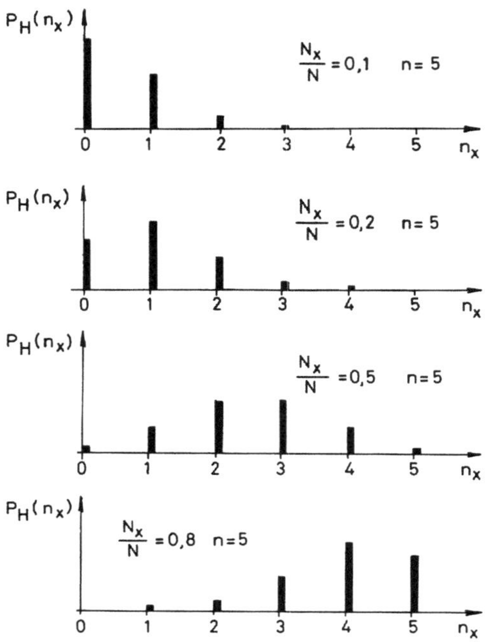

Figure 3.2.1 Hypergeometric distribution for different N_x/N with a sample size of n = 5 elements

After applying the rules for calculating with binomial coefficients:[4)]

$$\sum_{n_x=0}^{n} P_H(n_x) = \sum_{n_x=0}^{n} \frac{\binom{N_x}{n_x}\binom{N-N_x}{n-n_x}}{\binom{N}{n}} = \frac{\binom{N}{n}}{\binom{N}{n}} = 1 \qquad (3.2.7)$$

The distribution $P_H(n_x)$ is normalized.

The expectation $E(n_x)$ is given by:

$$E(n_x) = \sum_{n_x=0}^{n} n_x \cdot P_H(n_x) = \sum_{n_x=0}^{n} \frac{n_x \binom{N_x}{n_x}\binom{N-N_x}{n-N_x}}{\binom{N}{n}}$$

$$E(n_x) = \sum_{n_x=0}^{n} \frac{N_x \binom{N_x-1}{n_x-1}\binom{(N-1)-(N_x-1)}{(n-1)-(N_x-1)}}{\frac{N}{n}\binom{N-1}{n-1}}$$

$$\boxed{E(n_x) = n \cdot \frac{N_x}{N} = n \cdot p = \mu} \qquad (3.2.8)$$

where $p = N_x/N$ is the numeric concentration of the white elements in the population.

The variance is calculated as follows:

$$\sigma^2(n_x) = E(n_x^{\ 2}) - E^2(n_x) = E(n_x^{\ 2} - n_x) - E^2(n_x) + E(n_x)$$

$$\sigma^2(n_x) = E(n_x\ (n_x-1)) - n\frac{N_x}{N}\left(n\frac{N_x}{N} - 1\right)$$

$$= \sum_{n_x=0}^{n} \frac{n_x(n_x-1)\binom{N_x}{n_x}\binom{N-N_x}{n-N_x}}{\binom{N}{n}} - n\frac{N_x}{N}\left(n\frac{N_x}{N} - 1\right)$$

$$= \frac{N_x\ (N_x - 1)}{N\ (N-1)} \cdot n\ (n-1) - n\frac{N_x}{N}\left(n\frac{N_x}{N} - 1\right)$$

$$= \frac{n \cdot p}{N-1}\left\{ (N_x - 1)(n-1) - \left(n\frac{N_x}{N} - 1\right)(N-1) \right\}$$

$$\sigma^2(n_x) = n \cdot p\,(1 - p) \cdot \frac{N-n}{N-1} \qquad (3.2.9)$$

In contrast to the case of the expectation $E(n_x)$, the variance is dependent on the size N of the population (Figure 3.2.2)

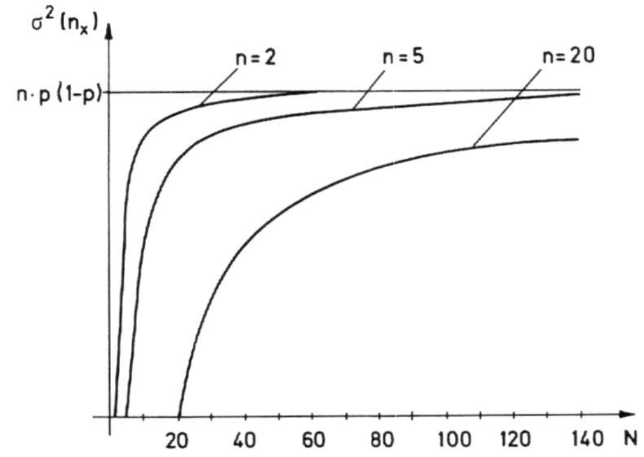

Figure 3.2.2 Dependence of the variance of the hypergeometric distribution on the size of the population N for a sample size of n = 2, n = 5 and n = 20

If N is increased beyond all limits, that is to say the population becomes infinitely large, the variance will tend towards a constant value:

$$\lim_{N\to\infty} \sigma^2(n_x) = n\,p\,(1 - p) \qquad (3.2.10)$$

The factor (N-n)/(N-1) is thus the measure of the "finiteness" of the population. Therefore it is also referred to as the finite population correction factor. For sufficiently large N the often used approximation holds:

$$\frac{N-n}{N-1} \approx \left(1 - \frac{n}{N}\right) \qquad (3.2.11)$$

3.3 Random sampling with replacement

3.3.1 Binomial distribution

If a sample of extent n is removed from an urn (population) of N elements, of which N_x are white and N_y are black, but each individual element is immediately replaced, then the probabilities $P_B(n_x)$ of obtaining exactly n_x white elements form a binomial distribution. The probability P(w) and P(b) of removing a white (w) or a black (b) from this urn, is equally large for each element and is equal to the proportion in the population:

$$P(w) = \frac{N_x}{N} = p$$
$$P(s) = \frac{N_y}{N} = q \qquad (3.3.1)$$

If an element is removed, its realization is recorded, the element is replaced and again an element is removed, then the following possibilities result for such a sample of two elements:

1st Draw	w	w	b	b
2nd Draw	w	b	w	b
Probability	p^2	p.q	q.p	q^2

If n elements are drawn from the urn under the same conditions, the resultant probability of first drawing n_x white and then $(n-n_x)$ black elements is:

$$P(\underbrace{w, w, w, \ldots}_{n_x}, \underbrace{b, b, b, \ldots}_{(n-n_x)}) = \underbrace{p.p.p\ldots}_{n_x} \underbrace{.q.q.q\ldots}_{(n-n_x)} = p^{n_x}.q^{n-n_x}$$

This is the probability for the arrangement taking account of the order. The requirement that n_x white and $(n-n_x)$ black elements are drawn, can, however, also be fulfilled in another order, for example:

$$P\ (w,s,w,w...,s,w,s,s...) = p^{n_x} \cdot q^{n-n_x}$$

This probability is of the same size. There are now exactly $\binom{n}{n_x}$ possibilities of arranging the elements in order and thus:

$$P_B\ (n_x) = \binom{n}{n_x}\ p^{n_x} \cdot q^{n-n_x} \qquad (3.3.2)$$

This is the binomial distribution. The binomial distribution is a discrete, one-dimensional probability distribution (Figure 3.3.1). In contrast to the hypergeometric distribution, it is independent of the size of the population.

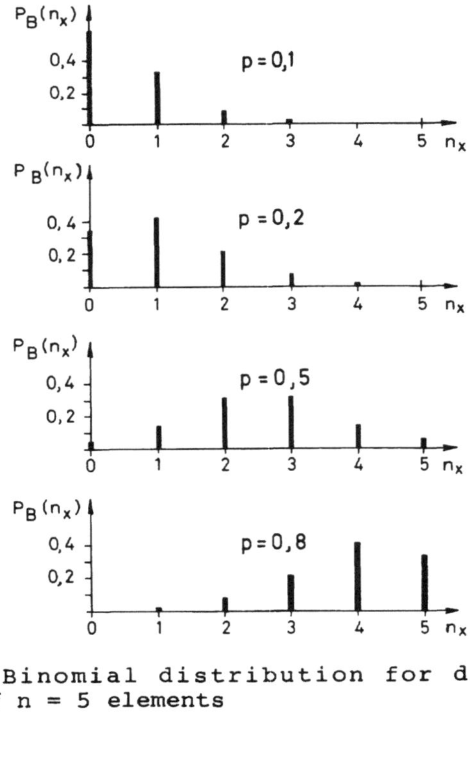

Figure 3.3.1 Binomial distribution for different p with sample size of n = 5 elements

The sum of the probabilities for all n_x is:

$$\sum_{n_x=0}^{n} P_B(n_x) = \sum_{n_x=0}^{n} \binom{n}{n_x} \cdot p^{n_x} \cdot q^{n-n_x} = (p+q)^n = 1 \quad (3.3.3)$$

The distribution is normalized. The expectation $E(n_x)$ is given by:

$$E(n_x) = \sum_{n_x=0}^{n} n_x \binom{n}{n_x} \cdot p^{n_x} q^{n-n_x} = \sum_{n_x=0}^{n} n \cdot \binom{n-1}{n_x-1} p^{n_x} \cdot q^{n-n_x}$$

$$= n \cdot p \sum_{n_x=0}^{n} \binom{n-1}{n_x-1} \cdot p^{n_x-1} q^{(n-1)-(n_x-1)} = n \cdot p\,(p+q)^{n-1}$$

$$\boxed{E(n_x) = n \cdot p = \mu} \quad (3.3.4)$$

The variance is calculated as follows:

$$\sigma^2(n_x) = E(n_x^2) - E^2(n_x) = E(n_x^2 - n_x) - E^2(n_x) + E(n_x)$$

$$= E(n_x(n_x-1)) - n \cdot p\,(n \cdot p - 1)$$

$$= \sum_{n_x=0}^{n} n_x(n_x-1) \cdot \binom{n}{n_x} \cdot p^{n_x} \cdot q^{n-n_x} - n \cdot p\,(n \cdot p - 1)$$

$$= n \cdot (n-1) \cdot p^2 \sum_{n_x=0}^{n} \binom{n-2}{n_x-2} \cdot p^{n_x-2} \cdot q^{(n-2)-(n_x-2)} - n \cdot p(n \cdot p - 1)$$

$$= n(n-1) \cdot p^2 - n \cdot p(n \cdot p - 1) = n \cdot p\,(p(n-1) - n \cdot p + 1)$$

$$\boxed{\sigma^2(n_x) = n \cdot p \cdot q} \quad (3.3.5)$$

The requirement that the probability of drawing an element of a certain realization from the urn is always the same, can also be approximately satisfied for samples without replacement, if the sample extent n is very small in comparison with the number N of the elements of the population. In this case the hypergeometric distribution $P_H(n_x)$ converges asymptotically to the binomial distribution, and the variance from equation (3.2.9) becomes that of equation (3.3.5) (Figure 3.2.2).

3.3.2 Poisson distribution

The Poisson distribution P_P is thought of as the distribution of "rare occurrences", ie. the probability of, for example, finding a white element in a population must be very small, $p \ll 1$. Furthermore, the sample extent n must be sufficiently large for the expectation $E(n_x) = n \cdot p = \mu$ to be a finite number, ie. $n \to \infty$ (Table 3.3.1). If additionally the assumptions of the binomial distribution obtain then:

$$P_B = \frac{n!}{n_x!\,(n-n_x)!} \left(\frac{\mu}{n}\right)^{n_x} \left(1 - \frac{\mu}{n}\right)^{n-n_x} \qquad (3.3.6)$$

$$P_P(n_x) = \lim_{\substack{n\to\infty \\ p\to 0}} P_B(n_x) = \frac{\mu^{n_x}}{n_x!} \lim_{n\to\infty} \left\{ \frac{n}{n}\,\frac{n-1}{n} \cdots \frac{n-(n_x-1)}{n} \cdot \left(1-\frac{\mu}{n}\right)^{n} \left(1-\frac{\mu}{n}\right)^{-n} \right\}$$

$$\boxed{P_P(n_x) = \frac{\mu^{n_x}}{n_x!} \cdot e^{-\mu}} \qquad (3.3.7)$$

This is the Poisson distribution. The Poisson distribution is a discrete, one-dimensional probability distribution (Figure 3.3.2). The expectation and the variance can be calculated from the corresponding equations of the binomial distribution by a limiting process for $p \to 0$, ie. $q \to 1$:

$$\boxed{E(n_x) = n \cdot p = \mu} \qquad (3.3.8)$$

und

$$\boxed{\sigma^2(n_x) = n \cdot p = \mu} \tag{3.3.9}$$

The variance of the Poisson distribution is identical with its expectation.

Table 3.3.1 Approximation to the binomial distribution by the Poisson distribution

n_x	Binomial distribution n = 4 p = 0,5	 n = 20 p = 0,1	 n = 100 p = 0,02	Poisson distribution $\mu = n \cdot p = 2$
0	0,06	0,12	0,13	0,14
1	0,25	0,27	0,27	0,27
2	0,38	0,29	0,27	0,27
3	0,25	0,19	0,18	0,18
4	0,06	0,09	0,09	0,09
5	0,00	0,03	0,04	0,04

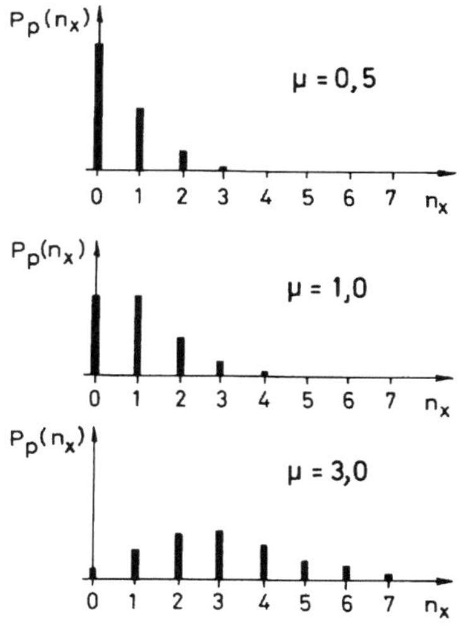

Figure 3.3.2 Poisson distribution for various expectations μ

3.3.3 Normal distribution

If, in addition to the assumptions of the binomial distribution, it is assumed that the sample extent n is very large, then the binomial distribution can be apporoximated by normal distribution with the mean $\mu = n \cdot p$ and the variance $\sigma^2 = n.p.q$:[5)]

$$h(n_x) = \frac{1}{\sigma\sqrt{2\pi}} \cdot e^{-\frac{1}{2}\left(\frac{n_x-\mu}{\sigma}\right)^2} \qquad (3.3.10)$$

For the values $n_x = 0, 1, \ldots n$, $P_B(n_x) \cong h(n_x)$.
The normal distribution was introduced by Gauss when working on the theory of measurement error. It is the most important continuous probability distribution, for the following reasons:

1) Many random variables are normally distributed.

2) Many random variables are approximately normally distributed. If the measured or presumed distributions are unimodal and are not too asymmetrical, then the assumption of a normal distribution will usually yield sensible and, in practice, useful results.

3) Quantities not distributed normally can frequently be given a normal distribution by transformation of the abscissa.

4) Under certain conditions, not specified in detail here, every sum of independent quantities $x = \Sigma x_i$ for $n \to \infty$ is asymptotically normally distributed (central limit theorem).

The distribution $h(n_x)$, the so-called bell-shaped curve is symmetrical with respect to the parameter μ. If the following substitution is carried out in equation (3.3.10)

$$\frac{n_x - \mu}{\sigma} = u, \qquad (3.3.11)$$

the standard normal distribution is obtained (Figure 3.3.3)

$$\boxed{h(u) = \frac{1}{\sqrt{2\pi}} \cdot e^{-\frac{1}{2}u^2}} \qquad (3.3.12)$$

The integral of the standard normal distribution over all n is:

$$\int_{-\infty}^{+\infty} h(u)\, du = \frac{1}{\sqrt{2\pi}} \cdot \int_{-\infty}^{+\infty} e^{-\frac{1}{2}u^2} \quad du = 1 \qquad (3.3.13)$$

Normal distribution h(u) is normalized.

The expectation E(u) is given by:

$$E(u) = \frac{1}{\sqrt{2\pi}} \cdot \int_{-\infty}^{+\infty} u \cdot e^{-\frac{1}{2}u^2} \quad du = 0 \qquad (3.3.14)$$

Therefore, using equation (3.3.11):

$$E\left(\frac{n_x - \mu}{\sigma}\right) = 0$$

$$\boxed{E(n_x) = \mu} \qquad (3.3.15)$$

The variance is given by:

$$\sigma^2(u) = \frac{1}{\sqrt{2\pi}} \cdot \int_{-\infty}^{+\infty} u^2 \cdot e^{-\frac{1}{2}u^2} = 1 \qquad (3.3.16)$$

It follows that:

$$\sigma^2\left(\frac{n_x - \mu}{\sigma}\right) = 1$$

$$\boxed{\sigma^2(n_x) = \sigma^2} \qquad (3.3.17)$$

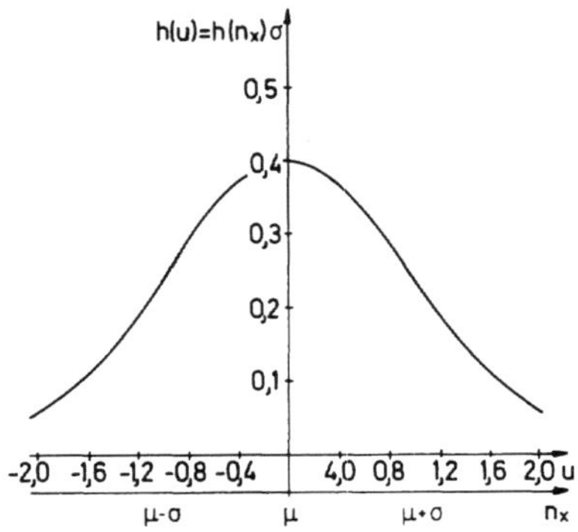

Figure 3.3.3 Standard normal distribution

Notes on the normal distribution (Figure 3.3.3)

1) The normal distribution is symmetrical with respect to its expectation μ.

2) The curve has a maximum at the position of the expectation μ:

$$\frac{d\,h(u)}{du} = -u \cdot \frac{1}{\sqrt{2\pi}} \cdot e^{-\frac{1}{2}u^2} = 0 \text{ für } u = 0$$

3) The normal distribution has a point of inflection at the positions $u = \pm 1$ or $x = \mu \pm \sigma$:

$$\frac{d^2h(u)}{du} = u^2 \cdot \frac{1}{\sqrt{2\pi}} \cdot e^{-\frac{1}{2}u^2} - \frac{1}{\sqrt{2\pi}} \cdot e^{-\frac{1}{2}u^2} = 0$$

$$u^2 - 1 = 0 \qquad u = \pm 1$$

4) The normal distribution is an approximation for the binomial distribution (Table 3.3.2). The distributions agree exactly only for $n \to \infty$. In practice, however, a useful approximation is already

Table 3.3.2 Normal distribution as an approximation to the binomial distribution

n_x	n = 5 binomial	p = 0,2 normal	n = 30 binomial	p = 0,2 normal	n = 5 binomial	p = 0,5 normal	n = 20 binomial	p = 0,5 normal
0	0,33	0,24	0,00	0,00	0,03	0,03	0,00	0,00
1	0,41	0,45	0,01	0,01	0,16	0,15	0,00	0,00
2	0,20	0,24	0,03	0,03	0,31	0,32	0,00	0,00
3	0,05	0,04	0,08	0,07	0,31	0,32	0,00	0,00
4	0,01	0,00	0,13	0,12	0,16	0,15	0,00	0,00
5	0,00	0,00	0,17	0,16	0,03	0,03	0,01	0,01
6	-		0,18	0,18	-		0,04	0,04
7	-		0,15	0,16	-		0,07	0,07
8	-		0,11	0,12	-		0,12	0,12
9	-		0,07	0,07	-		0,16	0,16
10	-		0,04	0,03	-		0,18	0,18

obtained for relatively small sample sizes. The closer p is to 0.5, the better is the approximation. Rule of thumb: in practical cases, the normal distribution can be used as an approximation, if $n.p > 4$ and $n.q > 4$.

The cumulative distribution function of the normal distribution is given by:

$$H(n_x) = \frac{1}{\sigma \sqrt{2\pi}} \int_{-\infty}^{n_x} e^{-\frac{1}{2}\left(\frac{\xi - u}{\sigma}\right)^2} \xi \qquad (3.3.18)$$

or in the standard form:

$$H(u) = \frac{1}{\sqrt{2\pi}} \cdot \int_{-\infty}^{u} e^{-\frac{1}{2}\xi^2} d\xi \qquad (3.3.19)$$

The normal distribution has been tabled in this form. According to section 2.4 the cumulative function is equal to the probability that the realization of the variables is smaller than or equal to a given quantity u or n_x.

Thus, the probability $P\ (a < u \leq b)$ that u is in the interval between a and b is given by:

$$P\ (a < u \leq b) = H(b) - H(a) \qquad (3.3.20)$$

Because of the symmetry of the distribution curve, the special case $a = -b$ gives:

$$P\ (-b < u \leq b) = H(b) - H(-b) = 2\,H(b) - 1 \qquad (3.3.21)$$

From this some characteristic values result:

a) $P(\mu - \sigma < n_x \leq \mu + \sigma) \approx 68\ \%$

b) $P(\mu - 2\sigma < n_x \leq \mu + 2\sigma) \approx 95{,}5\ \%$ (3.3.22)

c) $P(\mu - 3\sigma < n_x \leq \mu + 3\sigma) \approx 99{,}7\ \%$

The following formulae, also important for statistics, are obtained in a very similar way:

a) $P(\mu - 1{,}96\,\sigma < n_x \leq \mu + 1{,}96\,\sigma) = 95\ \%$

b) $P(\mu - 2{,}58\,\sigma < n_x \leq \mu + 2{,}58\,\sigma) = 99\ \%$ (3.3.23)

c) $P(\mu - 3{,}29\,\sigma < n_x \leq \mu + 3{,}29\,\sigma) = 99{,}9\ \%$

4 Sampling from a Population Having an Arbitrary Distribution of the Attribute

4.1 Sampling with replacement

In the previously considered cases the realizations of the distribution attribute of the elements were subject to a "black-white" decision. However, very frequently the elements of a population are characterized by a measurable attribute x, for example by size, by weight or concentration. The attribute x in the population has a distribution with the expectation $E(x) = \mu$ and the variance σ^2. Here it is not important whether the gradation of the attribute is continuous, as in the abovementioned examples, or whether it is discrete, such as, for example, in the case of prices and salaries where the smallest possible step is 0.01 $.

If a sample of size n is taken from this population of N elements, the attributes of the individual elements in the sample have the values x_1, x_2 ... x_n. A parameter of the total sample is the arithmetic mean of all elements in the sample:

$$\bar{x} = \frac{1}{n} \cdot \sum_{i=1}^{n} x_i \qquad (4.1.1)$$

If there are several samples, there is an x_1 for the 1st sample, an x_1 for the 2nd sample etc. and, analogously, an x_2 for the 1st sample, an x_2 for the 2nd sample etc. There is therefore a sequence of x_1, a sequence of x_2, ..., and a sequence of x_n values. Each of these sequences has a probability distribution with an expectation $E(x_i)$ and a variance $\sigma^2(x_i)$. If these samples are random samples, then $E(x_i) = \mu$ and $\sigma^2(x_i) = \sigma^2$.

The arithmetic means x also have a probability distribution with the expectation:

$$E(\bar{x}) = E\left(\frac{1}{n}\sum_{i=1}^{n} x_i\right) = \frac{1}{n}\cdot\sum_{i=1}^{n} E(x_i) = \mu \qquad (4.1.2)$$

The expectation of the sample mean agrees with the expectation μ of the population. This can also be expressed by saying that the arithmetic mean of the elements is unibased ("true to expectation").

If, as assumed, the samples are taken "with replacement" of the elements, then the values x_i are independent of one another. However, for independent quantities the variance of the sample mean is given by, according to equation (2.9.10):

$$\sigma^2(\bar{x}) = \sigma^2\left(\frac{1}{n}\sum_{i=1}^{n} x_i\right) = \frac{1}{n^2}\cdot\sum_{i=1}^{n}\sigma^2(x_i) = \frac{\sigma^2}{n} \qquad (4.1.3)$$

The variance of the sample mean $\sigma^2(\bar{x})$ is inversely proportional to the sample size.

4.2 Sampling without replacement

In the casė of sampling without replacing the individual elements the composition of the population changes with each element drawn. The probability of obtaining the realization x_i is influenced by all the elements already removed. The realizations x_i are therefore no longer independent. This finding does not alter the expectation of the sample mean:

$$E(\bar{x}) = E\left(\frac{1}{n}\sum_{i=1}^{n} x_i\right) = \frac{1}{n}\cdot\sum_{i=1}^{n} E(x_i) = \mu \qquad (4.2.1)$$

For samples without replacement of the individual elements, the expectation of the sample mean also agrees with the expectation of the population, in other words it is unbiased. However, for the variance, equation (4.1.3) does not hold, because there the independence of the realization x_i was assumed.

According to equation (2.5.10):

$$\sigma^2(\bar{x}) = E(\bar{x}^2) - E^2(\bar{x})$$

$$\sigma^2(\bar{x}) = E\left(\left(\frac{1}{n} \cdot \sum_{i=1}^{n} x_i\right)^2\right) - \mu^2 \qquad (4.2.2)$$

the square of the sum is expanded with the aid of the binomial theorem:

$$\sigma^2(\bar{x}) = \frac{1}{n^2}\, E\left(\sum_{i=1}^{n} x_i^2 + \sum_{i \neq j}^{n(n-1)} x_i \cdot x_j\right) - \mu^2$$

$$\sigma^2(\bar{x}) = \frac{1}{n^2}\left(E\left(\sum_{i=1}^{n} x_i^2\right) + \sum_{i \neq j}^{n(n-1)} E(x_i \cdot x_j) \right) - \mu^2$$

$$\sigma^2(\bar{x}) = \frac{E(x_i^2)}{n} + \frac{(n-1)}{n} \underset{i \neq j}{E} (x_i \cdot x_j) - \mu^2 \qquad (4.2.3)$$

The expectations $E(x_i^2)$ and $E_{i \neq j}(x_i \cdot x_j)$ are also expectations of the population and can be developed according to equations (2.9.2) and (2.9.7) respectively:

$$E(x_i^2) = \sigma^2 + \mu^2 \qquad (4.2.4)$$

$$\underset{i \neq j}{E} (x_i \cdot x_j) = \mathrm{Cov}\,(x_i, x_j) + \mu^2 \qquad (4.2.5)$$

Therefore:

$$\boxed{\sigma^2(\bar{x}) = \frac{\sigma^2}{n} + \frac{n-1}{n} \text{ Cov } (x_i, x_j)} \qquad (4.2.6)$$

If the realizations x_i and x_j are independent, the covariance disappears and equation (4.2.6) becomes equation (4.1.3).

In the case of sampling without replacement the covariance must be calculated. The arithmetic mean of the product $x_i \cdot x_j$ of the population is determined for this purpose:

$$\underset{i \neq j}{E} (x_i \cdot x_j) = \frac{1}{N(N-1)} \cdot \sum_{i \neq j}^{N(N-1)} x_i \cdot x_j \qquad (4.2.7)$$

Then, with the aid of the binomial theorem:

$$\underset{i \neq j}{E} (x_i \cdot x_j) = \frac{1}{N(N-1)} \left(\sum_{i=1}^{N} x_i^2 + \sum_{i \neq j}^{N(N-1)} x_i x_j - \sum_{i=1}^{N} x_i^2 \right)$$

$$\underset{i \neq j}{E} (x_i x_j) = \frac{1}{N(N-1)} \left(\left(\sum_{i=1}^{N} x_i \right)^2 - \sum_{i=1}^{N} x_i^2 \right)$$

$$\underset{i \neq j}{E} (x_i x_j) = \frac{1}{N(N-1)} \left(N^2 \left(\frac{1}{N} \sum_{i=1}^{N} x_i \right)^2 - N \left(\frac{1}{N} \sum_{i=1}^{N} x_i^2 \right) \right) \qquad (4.2.8)$$

The arithmetic means of the population agree with the expectations and therefore:

$$\underset{i \neq j}{E} (x_i \cdot x_j) = \frac{1}{N(N-1)} \left(N^2 \cdot \mu^2 - N \cdot E(x^2) \right) \qquad (4.2.9)$$

Using the formula (4.2.4) and substituting it in equation (4.2.5) leads to:

$$\mathrm{Cov}\,(x_i, x_j) = -\frac{1}{N-1}\cdot\sigma^2 \qquad (4.2.10)$$

From equation (4.2.6) the variance of the sample mean in the case of sampling "without replacement" can therefore be calculated:

$$\sigma^2(\bar{x}) = \frac{\sigma^2}{n} - \frac{n-1}{n(N-1)}\cdot\sigma^2$$

$$\boxed{\sigma^2(\bar{x}) = \frac{N-n}{N-1}\cdot\frac{\sigma^2}{n}} \qquad (4.2.11)$$

In comparison with the variance with replacement the finite population correction factor $(N-n)/(N-1)$, already defined in section 3.2, appears. If the population is very large compared with the sample size n, the dependence of the attribute realizations disappears, ie. the covariance becomes zero, and the variance agrees with the variance obtained for sampling with replacement. If the sample size is just $n = 1$, the two variances agree for all N.

The equations (4.1.3) and (4.2.11) permit determination of the sample size n for a known population (N, σ^2, μ) and a given error $\sigma^2(\bar{x})$.

5 Inference from the Sample About the Population (Confidence Intervals)

5.1 Sampling when the variance of the population is known

In section 4, the connection between the parameters $\sigma^2(x)$ and $E(x)$ of the population and $\sigma^2(x)$ and $E(\overline{x})$ of the distribution of the sample mean was derived. The parameters $\sigma^2(\bar{x})$ and $E(\bar{x})$ are however only theoretical values and have to be determined under the condition of an infinite number of samples. However, in practice an inference about a population is based on a single sample with the sample mean $\bar{x}$. This inference depends of course on chance factors during sampling. In the case of just a single sample, assertions about the connection between $\bar{x}$ and the expectation μ of the population can therefore only be made with a certain degree of confidence. The range of variation to be expected is also called a confidence interval. This problem can be solved analytically exactly only for certain distribution types (for example the normal distribution) of the population. If, for the population distribution only the variance σ^2 is known, but not the distribution type, then the Chebyshev inequality, given below without proof[6], supplies an upper estimate:

The probability P that the standardized quantity $|x - E(x)|/\sigma(x)$ is smaller than or equal to a given bound is greater than $(1-1/\varepsilon^2)$

$$P\left(\frac{|\bar{x}-E(\bar{x})|}{\sigma(\bar{x})} \le \varepsilon\right) \ge \left(1 - \frac{1}{\varepsilon^2}\right) \qquad (5.1.1)$$

For $E(\bar{x}) = E(x) = \mu$:

$$P\left(\frac{|\bar{x}-\mu|}{\sigma(\bar{x})} \le \varepsilon\right) \ge \left(1 - \frac{1}{\varepsilon^2}\right) \qquad (5.1.2)$$

If the bound $\varepsilon = 1$, 2 or 3, then the following relationships hold:

$$\text{a)}\quad P(\mu - \sigma(\bar{x}) \le \bar{x} \le \mu + \sigma(\bar{x})) \ge 0$$

$$\text{b)}\quad P(\mu - 2\,\sigma(\bar{x}) \le \bar{x} \le \mu + 2\,\sigma(\bar{x})) \ge 0{,}75 \qquad (5.1.3)$$

$$\text{c)}\quad P(\mu - 3\,\sigma(\bar{x}) \le \bar{x} \le \mu + 3\,\sigma(\bar{x})) \ge 0{,}889$$

In practice, a more frequent case is that the probability P is given and the size of the interval has to be determined for which the true value μ can be stated, with certainty P, to be lying within it. According to the inequality (5.1.2):

$$\frac{1}{\varepsilon^2} \ge 1 - P$$

$$\varepsilon \le \frac{1}{\sqrt{1-P}} \qquad (5.1.4)$$

Hence the probability is at least P that the true value lies within the interval

$$\boxed{\bar{x} - \frac{\sigma(\bar{x})}{\sqrt{1-P}} \le \mu \le \bar{x} + \frac{\sigma(\bar{x})}{\sqrt{1-P}}} \qquad (5.1.5)$$

The interval is called a confidence interval.

If the sample consists of n elements, then, according to equation (4.1.3), for sampling "with replacement":

$$\boxed{\bar{x} - \frac{\sigma}{\sqrt{n}}\,\frac{1}{\sqrt{1-P}} \le \mu \le \bar{x} + \frac{\sigma}{\sqrt{n}} \cdot \frac{1}{\sqrt{1-P}}} \qquad (5.1.6)$$

and, according to equation (4.2.11) for sampling without replacement:

$$\bar{x} - \frac{\sigma}{\sqrt{n}}\sqrt{\frac{N-n}{N-1}} \cdot \frac{1}{\sqrt{1-P}} \leq \mu \leq \bar{x} + \frac{\sigma}{\sqrt{n}}\sqrt{\frac{N-n}{N-1}} \frac{1}{\sqrt{1-P}} \tag{5.1.7}$$

The use of equations (5.1.6) and (5.1.7) presumes knowledge of the variance σ^2 of the population. However, in most cases this is also not known. In this case the variance $\sigma^2(\bar{x})$ in equation (5.1.5) must be replaced by the sample variance

$$s^2(x) = \frac{1}{n-1} \sum_{i=1}^{n} (x_i - \bar{x})^2 \tag{5.1.8}$$

In practice this usually leads to usable approximations, though it is not quite correct mathematically. In the case of neither the expectation μ nor the variance σ^2 of the population being known, mathematically exact statements can be made only for normally distributed populations.

5.2. Normally distributed populations with known variance σ^2

It was shown in section 3.3.3 that in the case of large sample sizes n, but which are still small in comparison with the number of elements in the population, the distribution of the sample attribute approximates very accurately to the normal distribution. In general this also holds for the case in which the population has an arbitrary distribution with the expectation $E(x) = \mu$ and the variance $\sigma^2(x) = \sigma^2$ (central limit theorem).
Consequently, the sample mean is also distributed approximately normally with the expectation $E(\bar{x}) = \mu$ and the variance $\sigma^2(\bar{x})$. If the population itself is normally distributed, this statement holds exactly. Instead of the inequality (5.1.2), the condition according to equation (3.3.21), with the standardization $u = (\bar{x} -$

$\mu)/\sigma(x)$, holds:

$$P\left(\frac{|(\bar{x}-\mu)|}{\sigma(\bar{x})} \leq c\right) = 2\,H(c) - 1 \qquad (5.2.1)$$

H(c) is the tabulated, standard normal distribution. If the bounds c = 1, 2 or 3 are chosen, then:

a) $P(\mu - \sigma(\bar{x}) < \bar{x} \leq \mu + \sigma(\bar{x})\ ) = 0,68$

b) $P(\mu - 2\,\sigma(\bar{x}) < \bar{x} \leq \mu + 2\,\sigma(\mathbf{x}) = 0,955$ (5.2.2)

c) $P(\mu - 3\,\sigma\,(\bar{x}) < \bar{x} \leq \mu + 3\,\sigma(\bar{x}) = 0,999$

The equations (5.2.2 a, b and c) satisfy the inequalities (5.1.3 a, b and c). However, the use of the normal distribution results in considerably more precise statements.

If the probability P is given, then the true value μ lies, with the probability P, within the confidence interval:

$$\bar{x} - c \cdot \sigma(\bar{x}) \leq \mu < \bar{x} + c \cdot \sigma(\bar{x}) \qquad (5.2.3)$$

c is determined for a known P from the normal distribution H:

$$S = P = 2\,H(c) - 1 \qquad (5.2.4)$$

If the sample consists of n elements, then, according to equations (4.1.3) and (4.2.11), for sampling with replacement:

$$\boxed{\bar{x} - c\,\frac{\sigma}{\sqrt{n}} \leq \mu < \bar{x} + c \cdot \frac{\sigma}{\sqrt{n}}} \qquad (5.2.5)$$

and for sampling without replacement:

$$\bar{x} - c\frac{\sigma}{\sqrt{n}}\sqrt{\frac{N-n}{N-1}} \le \mu < \bar{x} + c\frac{\sigma}{\sqrt{n}}\sqrt{\frac{N-n}{N-1}} \qquad (5.2.6)$$

σ is the known standard deviation of the population.

Example:

A sample of $n = 3$ particles with metal contents $x_1 = 46$ %, $x_2 = 49$ % and $x_3 = 52$ % was taken from a sieved off, even-grained fraction of an ore. What is the size of the 95 % confidence interval for the mean x - (a) for an unknown distribution of the population with $\sigma^2 = 25 \times 10^{-4}$ or (b) for an assumed normal distribution with $\sigma^2 = 25 \times 10^{-4}$?

$\bar{x} = \frac{1}{3}\sum_{i=1}^{3} x_i = 0{,}49 \qquad S = P = 0{,}95 \qquad \sigma = 5 : 10^{-2} \qquad n = 3$

a) With equation (5.1.6):

$$0{,}49 - 0{,}13 \le \mu < 0{,}49 + 0{,}13$$

$$36\ \% \le \mu < 62\ \%$$

b) For P = 0,95, c = 1,960
With equation (5.2.5)

$$0{,}49 - 0{,}06 \le \mu < 0{,}49 + 0{,}06$$

$$43\ \% \le \mu < 55\ \%$$

5.3 Sampling from normally distributed populations with unknown variance

5.3.1 Chi-square distribution (confidence interval for sample variances)

It is assumed that the values x of the population are normally distributed. The sample mean will therefore also be exactly normally distributed, with the same expectation $E(\bar{x}) = \mu$, and the variance from equation (4.1.3) or (4.2.11). Very frequently the variance of the population is not known. In this case, the assumption of a normally distributed population yields exact results. It is true without proof:[7] If the attribute x of the population is normally distributed with the variance σ^2 and the expectation μ, and if, using equation (2.3.2), the sample variance is defined by

$$s^2 = \frac{1}{n-1} \sum_{j=1}^{n} (x_j - \bar{x})^2 \qquad (5.3.1)$$

then the quantity

$$\chi^2 = (n-1) \cdot \frac{s^2}{\sigma^2} \qquad (5.3.2)$$

is chi-square distributed with the degree of freedom f = n - 1:

$$\boxed{f_f(\chi^2) = \frac{1}{2^{f/2}\,\Gamma(\frac{f}{2})} \cdot (\chi^2)^{\frac{f-2}{2}} \, e^{-\frac{\chi^2}{2}}} \qquad (5.3.3)$$

The chi-square distribution is a one dimensional, nonsymmetrical density distribution (Figure 5.3.1).

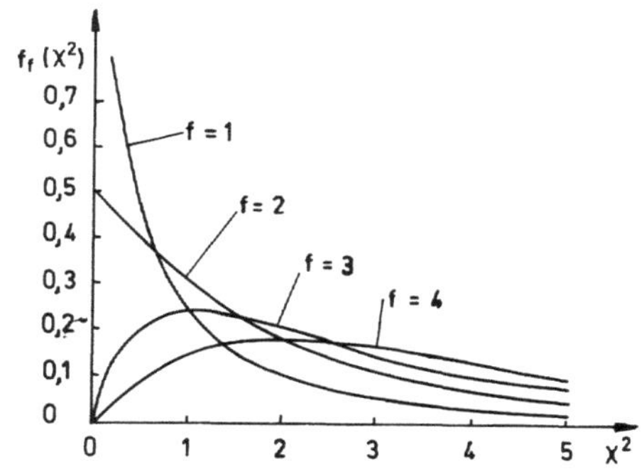

Figure 5.3.1 Chi-square density distribution for various degrees of freedom f

The distribution sum has been tabulated[8]:

$$F_f(\chi^2) = \frac{1}{2^{f/2}\,\Gamma(\frac{f}{2})} \int_0^{\chi^2} \xi^{\frac{f-2}{2}} \cdot e^{-\frac{\xi^2}{2}} \, d\xi \qquad (5.3.4)$$

The expectation is given by:

$$E_f(\chi^2) = n - 1 = f \qquad (5.3.5)$$

and the variance is given by:

$$\sigma_f^2 (\chi^2) = 2\,(n-1) = 2\,f \qquad (5.3.6)$$

The probability P that the quantity $\chi^2 = (n-1)s^2/\sigma^2$ is within certain limits χ_e^2 and χ_u^2 is:

$$P\,(\chi_u^2 < \chi^2 \leq \chi_o^2) = F_f(\chi_o^2) - F_f(\chi_u^2) \qquad (5.3.7)$$

If the probability P is given, then the quantity σ^2 is within the limits:

$$\frac{(n-1)\cdot s^2}{\chi_o^2} \leq \sigma^2 < \frac{(n-1)\cdot s^2}{\chi_u^2} \qquad (5.3.8)$$

The limits can be calculated from the required probability. The nonsymmetrical chi-square distribution results in correspondingly nonsymmetrical confidence regions. If the lower limit is set at $P_l = 0,025$ and the upper limit at $P_u = 0,975$, then, for example, for a sample size of n = 3: $\chi_e^2 = 0,05$ and $\chi_u^2 = 7,38$.

Example:

In the Example from section 5.2 with $x_1 = 0,46$, $x_2 = 0,49$, $x_3 = 0,52$ and n = 3 it followed that:

$s^2 = 9 \times 10^{-4}$

Equation (5.3.8) gives, for 95 % probability $2.4 \times 10^{-4} < 360 \times 10^{-4}$.

This interval is consistent with the given quantity $\sigma^2 = 25 \times 10^{-4}$.

5.3.2 The t distribution (confidence intervals for means when the variance is unknown)

In the preceding investigation, the confidence intervals for means were always based on the standardized quantity $(\bar{x} - E(\bar{x}))/\sigma(\bar{x})$. However, this makes sense only if the variance $\sigma^2(\bar{x})$ is known, otherwise $\sigma^2(\bar{x})$ must be replaced by the sample variance s^2. This is possible because of the σ^2 quantity (equations 5.3.2):

$$t = \frac{(\bar{x} - E(\bar{x}))/\sigma(\bar{x})}{\sqrt{\chi^2/(n-1)}} \qquad (5.3.9)$$

If the values x in the population are normally distributed, the quantity t has a distribution, named for the assumed name Student of the statistician W.S. Gosset, namely the Student distribution or t distribution:

$$f_f(t) = \frac{\Gamma\left(\frac{f+1}{2}\right)}{\sqrt{f \cdot \pi}\;\Gamma\left(\frac{f}{2}\right)} \quad \frac{1}{\left(1 + \frac{t^2}{f}\right)^{\frac{f+1}{2}}} \tag{5.3.10}$$

f = n - 1 is the degree of freedom and a parameter of the t distribution. The t distribution is a one-dimensional symmetrical density distribution (Figure 5.3.2).

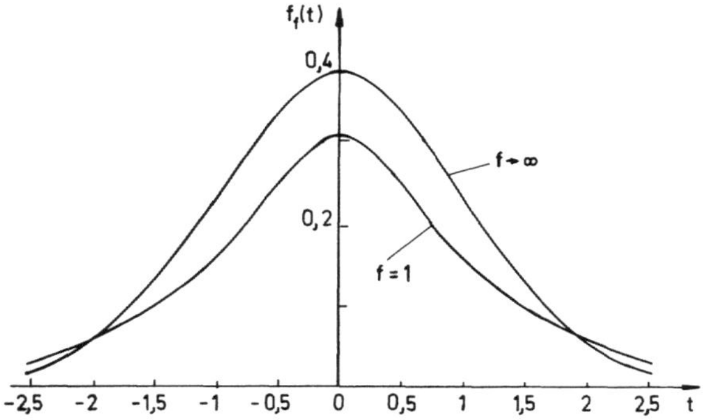

Figure 5.3.2 Density of the t distribution for degrees of freedom f = 1 and f = ∞

The distribution sum $F_f(t)$ has been tabulated[8].

$$F_f(t) = \frac{\Gamma\left(\frac{f+1}{2}\right)}{\sqrt{f \cdot \pi}\;\Gamma\left(\frac{f}{2}\right)} \int_{-\infty}^{t} \frac{d\xi}{\left(1 + \frac{\xi^2}{f}\right)^{\frac{f+1}{2}}} \tag{5.3.11}$$

The expectation of the distribution is given by:

$$E(t) = 0 \qquad \text{für} \quad f > 1 \tag{5.3.12}$$

and the variance is given by:

$$\sigma^2(t) = \frac{f}{f-2} \qquad \text{für } f > 2 \tag{5.3.13}$$

Since the sample variance s^2 enters into the quantity t (equation 5.3.9), it must be the case that $n \geq 2$ and $f \geq 1$. In the limiting case when $f \to \infty$, the t distribution becomes the standard normal distribution (Table 5.3.1).

Table 5.3.1 Approximation of the t distribution by the normal distribution

t	f = 1	f = 2	f = 3	f = 9	normal f → ∞
0	0,32	0,35	0,37	0,39	0,40
0,5	0,25	0,30	0,31	0,34	0,35
1,0	0,16	0,19	0,21	0,23	0,24
1,5	0,10	0,11	0,12	0,13	0,13
2,0	0,06	0,07	0,07	0,06	0,05
2,5	0,04	0,04	0,04	0,03	0,02
3,0	0,03	0,03	0,02	0,01	0,00

The probability P that the quantity t lies within $\pm$ a is, because of the symmetry of the t distribution:

$$P(-a < t \leq +a) = F(a) - F(-a) = 2\,F(a) - 1 \tag{5.3.14}$$

If the probability P is given, then t has the probability P of lying within the interval:

$$- a < t \leq + a \qquad (5.3.15)$$

From equation (5.3.9) follows that:

$$t = \frac{(\bar{x}-\mu)}{\sigma(\bar{x})} \cdot \frac{\sigma}{s} = \sqrt{n}\ \frac{\bar{x}-\mu}{s}$$

and hence the true value μ lies with the probability P within the confidence interval

$$\boxed{\bar{x} - a \cdot \frac{s}{\sqrt{n}} \leq \mu < \bar{x} + a \cdot \frac{s}{\sqrt{n}}} \qquad (5.3.16)$$

the limits ± a can be calculated from the required probability; for example, for P = 0,95 and n = 3, a = 4,3.

<u>Example:</u>

In the example in Section 5.2 with

x_1 = 0,46 x_2 = 0,49 x_3 = 0,52 n = 3

$\bar{x}$ = 0,49 and $s^2 = 9 \cdot 10^{-4}$,

equation (5.3.16) gives, for 95 % probability:

42 % < μ ≦ 56 %

The interval differs only insignificantly from the known variance in Section 5.2.

6 Sampling Procedures

6.1 Random and systematic sampling

As a model for random selection, the urn scheme has already been developed for the derivation of the sampling distribution (Section 3.1). In industrial sampling, the urn scheme is not to be realized as such but only in a figurative sense. The so called random number tables (Tables 6.1.1) are frequently used for this purpose. Random number tables are, for example, contained in the books: Owen, O.B., Handbook of Statistical Tables. Reading, Mass.: Addison-Wesley 1962, Horton, H.B., Random Decimal Digits, Interstate Commerce Commission 1949, and Kreyszig, El, Statistische Methoden und ihre Anwendungen, Vandenhoeck u. Ruprecht, Göttingen 1968.

Random numbers are collated in such a manner that each numeral from 0 - 9 has the same probability of appearing as any digit in a number. Such a table is in a sense a supply of randomness. If sampling is to be based on random numbers, all elements of the population have first to be numbered from 1 to N, some number is then picked by chance from the table and other numbers are chosen by proceeding according to some selected rule. All numbers which are not larger than the total number N of the elements in the population are noted down until the disired sample size has been reached. Since the numerals 0 - 9 are equally distributed in a random number, only the number of digits desired in each case can be taken into account when multidigit random numbers are given (for example, the first two numerals of a fourdigit random number, if the sample size is 100 elements). If the same numbers occur two or more times and if the sample variable of this element is correspondingly noted down two or more times, then the elements of the population which correspond to the numbers noted down constitute a sample "with replacement". If all numbers which occur more than once are only noted down once, then the noted down elements constitute a sample "without replacement".

Table 6.1.1 Random numbers*)

6977	6081	6733	6363	7124	2985	3434	8499	1989	3109
8377	8357	3350	4595	6235	6532	6556	8575	3370	1992
3034	9586	1765	8717	2363	4741	8509	4710	4886	2410
9903	9539	5787	8692	3367	8343	0942	5605	4772	4438
6955	8569	2111	7416	8660	9795	6551	2171	4123	5869
5483	0587	8690	2422	7334	3626	6218	3210	6876	2500
5733	4729	1443	6895	7864	3421	3390	6435	2518	5483
0126	9533	3548	2999	0951	1381	6696	6250	9404	3552
4329	9158	9291	2629	1976	5815	9556	9016	6604	5456
3776	8729	0478	4410	0551	0223	4173	8312	7975	6768
1539	0850	5347	2268	5847	3227	0650	8474	5658	7783
3390	5370	0046	5861	5215	0102	1071	6404	9787	8271
1562	6106	5840	8594	8217	5062	0410	7008	1476	0788
9408	3412	3881	4737	9370	1603	0916	6167	4329	9370
2306	4439	5476	3383	8966	8757	0861	1202	8422	4241
8196	8288	9236	8022	1886	1765	8925	6413	5370	0463
8489	5702	8822	5071	8599	2016	3681	2403	6983	0307
7652	6009	5347	2476	2345	9456	0441	4013	1246	3582
2450	3068	3892	7924	4594	5814	9135	1562	9506	7492
1464	2104	2222	4195	5376	7292	0876	3923	1368	9830
9256	5105	3984	1032	5298	4652	2534	8515	7818	1676
2337	5302	3016	7027	4269	7610	0337	7981	9892	0878
6127	2754	9052	6676	7836	5739	7486	2727	9952	7943
7703	5246	5965	7505	7656	0439	3194	3642	1598	6388
2380	8220	0781	5001	5831	0052	9742	3222	4256	5206
4934	0027	0957	8223	8835	6847	4963	4948	2015	3262
5658	7890	9610	4052	2378	7462	4422	2014	2629	2152
6628	4078	1603	1126	4666	1626	1835	0553	1377	5172
4022	8875	8190	1670	2429	6103	4391	8594	8410	2939
6969	0067	2907	9407	8325	9885	6218	2993	6816	1394
1936	8890	0633	4732	3074	0701	7147	9311	9060	5571
7533	7325	5710	6848	5280	0586	8167	3573	6810	4675
8545	7774	9637	6347	3831	7486	1553	2762	0008	7850
9191	3756	1190	2500	1048	9191	3495	2218	0800	0224
9651	2710	6095	8724	9870	3558	7113	2313	6895	1360

*)Owen, D.B., Handbook of Statistical Tables. Reading, Mass.: Addison-Wesley 1962

The proposed method of random selection frequently requires too much effort in practice. It is accordingly preferable to take systematic samples. If a sample of n is to be taken from a population of N elements, each i-th element is selected, for example by sampling from a conveyor belt according to prescribed belt distances. In practice, systematic samples of that type are frequently treated as random samples. Strictly speaking, this is only correct, however, if the population is stochastically homogeneous, i.e. a random mixture. If this precondition is not satisfied, errors can result in systematic sampling:

1. If the attribute to be investigated has a periodic variation, for example because a component on the conveyor belt is metered-in with periodic variations, and if the distance between the sample elements coincides with the length of the variation, then the sample is no longer representative and a statement regarding the expectation of the attribute in the population is no longer possible with purely statistical aids.

2. If the disturbances are random and if the length of the variation is larger than the size of the individual elements, adjacent elements will no longer be independent of one another, in other words they will be correlated with one another. The size of the actually independent regions is of the same order as the variation length, expressed in terms of the correlation length. Stange (1958) was the first to undertake fundamental work on the correlation length on a conveyor belt and this work is discussed in detail in Sections 12 and 13. Despite the greater risk of error, systematic selection is extraordinarily frequently used because of its simplicity. Compared with systematic selection, working with random numbers usually requires considerably more effort.

6.2 Stratified sample (Table 6.2.1 and 6.2.2)

Stratified sampling is a process in which the population is divided into several classes each class is treated as a population in its own right and a sample is taken from each. In this process, the classes or strata may already be given by the state of the material or the mode of transport, for example in the form of barrels or bags, or they can be chosen by the sampler. An instance of the latter situation is the taking of a sample from a sedimentation cake in order to determine the particle size distribution. Sedimentation results in relatively more uniform strata with respect to particle size, which, on using stratified sampling, can lead to greater confidence. Here, the "stratification" of sampling is thus to be interpreted literally. In addition to the question about the degree of this confidence, there is still the problem of optimum strata selection. The terms which are required for the solution of this problem have been collated in Table (6.2.1).

If a random sample of n elements is taken without stratification from a population with the expectation μ and the variance σ^2, formulae (4.1.3) and (4.2.11), which were derived in section 4, hold, for example, for sampling without replacement:

$$E(\bar{x}) = \mu \qquad \text{(6.2.1) Table 6.2.1, row 7}$$

$$\sigma^2(\bar{x}) = \frac{N-n}{N-1} \cdot \frac{\sigma^2}{n} \qquad \text{(6.2.2) Table 6.2.1, row 8}$$

In the case of a stratified sample, the mean x_i is first calculated from the n_i elements of the i-th stratum:

$$\bar{x}_i = \frac{1}{n_i} \sum_{j=1}^{n_i} x_{ij} \qquad \text{(6.2.3) Table 6.2.1, row 9}$$

Since each stratum is considerd as a population on its own, an expectation can be defined for this stratum

$$E(x_{ij}|\,i) = \mu_i \qquad \text{(6.2.4) Table 6.2.1, row 9}$$

$E(x_{ij}|i)$ denotes, according to definition (Section 2.7), the conditional expectation of the quantity x_{ij}. The condition is

Table 6.2.1 Terms in stratified sampling

	Population	Sample
1 Number of elements	N	n
2 Number of strata	k	k
3 Serial numbers of the strata	$1 \le i \le k$	$1 \le i \le k$
4 Elements in the i-th stratum	N_i	n_i
5 Numbers of elements in the i-th stratum	$1 \le j \le N_i$	$1 \le j \le n_i$
6 Attribute of the elements	x_{ij}	x_{ij}
7 Expectation and mean without stratification	$E(x_{ij}) = \mu$	$\bar{x} = \frac{1}{n} \sum_{i=1}^{k} \sum_{j=1}^{n_i} x_{ij}$
8 Variances without stratification	$\sigma^2(x_{ij}) = \sigma^2$	$\sigma^2(\bar{x}) = \frac{N-n}{N-1} \cdot \frac{\sigma^2}{n}$
9 Expectation and mean of the i-th stratum	$E(x_{ij}\|i) = \mu_i$	$\bar{x}_i = \frac{1}{n_i} \sum_{j=1}^{n_i} x_{ij}$
10 Variances of the i-th stratum	$\sigma^2(x_{ij}\|i) = \sigma_i^2$	$\sigma^2(\bar{x}_i\|i) = \frac{N_i-n_i}{N_i-1} \cdot \frac{\sigma_i^2}{n_i}$
11 Expectation and mean between the strata	$E\{(E(x_{ij}\|i)\} = \mu$	$\bar{\bar{x}}_i = \bar{\bar{x}} = \frac{1}{N} \sum_{i=1}^{k} N_i \bar{x}_i$ $\bar{\bar{x}} = \frac{1}{N} \sum_{i=1}^{k} \left\{ N_i \frac{1}{n_i} \sum_{j=1}^{n_i} x_{ij} \right\}$
12 Variance for stratified sample	Equation 6.2.12	$\sigma^2(\bar{\bar{x}}) = \sum_{i=1}^{k} (\frac{N_i}{N})^2 \frac{N_i-n_i}{N_i-1} \cdot \frac{\sigma^2(x_{ij}\|i)}{n_i}$
13 Variance for proportional subdivision $\frac{n_i}{N_i} = \frac{n}{N}$	Equation 6.3.2b	$\sigma_p^2(\bar{\bar{x}}) = (\frac{1}{n} - \frac{1}{N}) \cdot E(\sigma_i^2)$
14 Var. for cost-optimized subdivision $\frac{n_i}{N_i} = \frac{n \cdot \sigma_i}{\sum_{i=1}^{k} N_i \cdot \sigma_i}$	Equation 6.4.12	$\sigma_k^2(\bar{\bar{x}}) = \frac{1}{n} E^2(\sigma_i) - \frac{1}{N} E(\sigma_i^2)$

Table 6.2.2 Stratified sampling

1st step	Selection of strata $i = 1, 2, \ldots, k$
2nd step	Fixing the size of the units to be investigated and hence determining N_i and $N = \Sigma N_i$
3rd step	Choice of sample sizes n_i
4th step	Determination of the variances $\sigma^2(x_{ij}\mid i) = \sigma^2_i$ within the strata by prior experiments or by estimation using the respective n_i measured values:

$$\sigma^2_i \approx \frac{1}{n_i-1} \sum_{j=1}^{n_i} (x_{ij}-\bar{x}_i)^2;\quad \bar{x}_i = \frac{1}{n_i} \sum_{j=1}^{n_i} x_{ij}$$

5th step Determination of the sample mean form the x_{ij}:

(6.2.7)
$$\bar{\bar{x}} = \frac{1}{N} \sum_{i=1}^{k} N_i \bar{x}_i = \frac{1}{N} \sum_{i=1}^{k} \left\{ N_i \cdot \frac{1}{n_i} \cdot \sum_{j=1}^{n_i} x_{ij} \right\}$$

6th step Determination of the variance of the sample means:

(6.2.12)
$$\sigma^2(\bar{\bar{x}}) = \sum_{i=1}^{k} \left(\frac{N_i}{N}\right)^2 \frac{N_i-n_i}{N_i-1} \frac{\sigma^2_i}{n_i}$$

7th step Preassigning a confidence coefficient S(e.q.95%).

8th step Find the solution c from the normal distribution*):

(5.2.4)
$$S = 2\,H(c) - 1$$

9th step Hence the confidence region with the confidence coefficient S and $n = \Sigma n_i$ is given by

(5.2.6)
$$\mu = \bar{\bar{x}} \pm c \frac{\sigma}{\sqrt{n}} \sqrt{\frac{N-n}{N-1}}$$

For $N \gg n$:

(5.2.5)
$$\mu = \bar{\bar{x}} \pm c \frac{\sigma}{\sqrt{n}}$$

*)It is assumed that the values are approximately normally distributed, otherwise equation (5.1.3) has to be used.

that in each case only the i-th stratum is considered. Correspondingly, there is also a conditional variance for the i-th stratum.

$$\sigma^2(x_{ij}|i) = \sigma_i^2 \qquad \text{(6.2.5) Table 6.2.1, row 10}$$

If the i-th stratum consists of N_i elements and if, as assumed, a sample of size n_i is taken without replacement, the conditional variance of the sample mean of the i-th stratum is given by:

$$\sigma^2(\bar{x}_i|i) = \frac{N_i - n_i}{N_i - 1} \; \frac{\sigma^2(x_{ij}|i)}{n_i} \qquad \text{(6.2.6) Table 6.2.1, row 10}$$

The mean of all $\bar{x}_i$, weighted according to the respective stratum size N_i, is considered to be the estimate of the stratified sample:

$$\bar{\bar{x}}_i = \bar{\bar{x}} = \frac{1}{N} \sum_{i=1}^{k} N_i \bar{x}_i = \frac{1}{N} \sum_{i=1}^{k} \left\{ N_i \cdot \frac{1}{n_i} \sum_{j=1}^{n_i} x_{ij} \right\}$$

(6.2.7) Table 6.2.1, row 11

It can be shown directly that $\bar{\bar{x}}$ is unbiased:

$$E(\bar{\bar{x}}) = \frac{1}{N} \sum_{i=1}^{k} \left\{ \frac{N_i}{n_i} \sum_{j=1}^{n_i} E(x_{ij}) \right\} \qquad (6.2.8)$$

In conjunction with equation (6.2.1) it follows that

$$E(\bar{\bar{x}}) = \frac{1}{N} \sum_{i=1}^{k} (N_i \cdot \mu_i) = \mu \qquad (6.2.9)$$

This result corresponds to the computational rule for conditional expectations, according to equation (2.9.16):

$$E \{ E(x_{ij} | i) \} = E(x_{ij}) = \mu \qquad \text{(6.2.10) Table 6.2.1, row 11}$$

The variance of the sample mean $\bar{\bar{x}}$ is given by:

$$\sigma^2(\bar{\bar{x}}) = \sigma^2 \left\{ \frac{1}{N} \sum_{i=1}^{k} \left(\frac{N_i}{n_i} \sum_{j=1}^{n_i} x_{ij} \right) \right\}$$

$$= \sigma^2 \left\{ \frac{1}{N} \sum_{i=1}^{k} (N_i \cdot \bar{x}_i) \right\}$$

Using the computational rules for variances:

$$\sigma^2(\bar{\bar{x}}) = \frac{1}{N^2} \sum_{i=1}^{k} N_i^{\,2} \cdot \sigma^2 (\bar{x}_i | i) \qquad (6.2.11)$$

Here it is assumed that the means $\bar{x}_i$ are independent of one another. In conjunction with equation (6.2.6), this leads to:

$$\boxed{\sigma^2(\bar{\bar{x}}) = \sum_{i=1}^{k} \left(\frac{N_i}{N} \right)^2 \frac{N_i - n_i}{N_i - 1} \cdot \frac{\sigma^2(x_{ij} | i)}{n_i}}$$

(6.2.12) Table 6.2.1, row 12

In order to calculate $\sigma^2(\bar{\bar{x}})$ the $\sigma^2(x_{ij} | i)$, that is to say the variances within the strata, must be known. In practice it is most frequently necessary to estimate these values from prior experiments, in order to gain indications regarding the size of the $\sigma^2(x_{ij} | i)$ and hence to calculate the variance $\sigma^2(\bar{\bar{x}})$ approximately.

For a given stratification and given total sample size n, the n sample elements can still be freely allocated to the individual strata.

6.3 Proportional subdivision a stratified sample (Table 6.3.1)

The simplest possibility of subdivision is to select from each stratum, in proportion to its size N_i, the size of the part sample n_i:

$$\frac{n_i}{N_i} = \frac{n}{N} = \text{konst} \qquad (6.3.1)$$

If, in addition, it is assumed that the individual strata are large, with $N_i \gg 1$, it follows from equation (6.3.1) that the variance for proportional subdivision is given by:

$$\sigma_p^2(\bar{\bar{x}}) = \sum_{i=1}^{k} \frac{N_i^2}{N^2}\left(1 - \frac{n_i}{N_i}\right) \cdot \frac{N_i}{n_i} \cdot \sigma^2(x_{ij}|i) \cdot \frac{1}{N}$$

$$\sigma_p^2(\bar{\bar{x}}) = \left(1 - \frac{n}{N}\right)\frac{1}{n}\sum_{i=1}^{k} \frac{N_i}{N} \cdot \sigma^2(x_{ij}|i)$$

$$\sigma_p^2(\bar{\bar{x}}) = \left(\frac{1}{n} - \frac{1}{N}\right) \cdot \frac{1}{N}\sum_{i=1}^{k} N_i \cdot \sigma^2(x_{ij}|i) \qquad (6.3.2a)$$

The mean of the $\sigma^2(x_{ij}|i) = \sigma^2_i$ is formed over all strata, and is thus a mean of the population and hence an expectation:

$$\boxed{\sigma_p^2(\bar{\bar{x}}) = \left(\frac{1}{n} - \frac{1}{N}\right) \cdot E(\sigma_i^2)} \qquad (6.3.2\text{ b}) \text{ Table 2.1 row 13}$$

A comparsion of the variance $\sigma_p^2(\bar{\bar{x}})$ for proportional subdivision with the variance without stratification can be readily carried out using the computational rules for conditional variances (equation 2.9.17). Then:

$$\sigma^2(x_{ij}) = \sigma^2 = E\left(\sigma^2(x_{ij}|i)\right) + \sigma^2\left(E(x_{ij}|i)\right) \qquad (6.3.3)$$

Table 6.3.1 Proportional subdivison for stratified sample

1st step	Selection of strata i = 1, 2, ... k
2nd step	Fixing the size of the investigation units and hence determining the N_i and $N = \Sigma N_i$
3rd step	Determining the sample sizes n_i by proportional subdivision for a chosen total sample size
(6.3.1)	$n_i = \frac{n}{N} \cdot N_i$
4th step	Finding the variances $\sigma^2(x_{ij} \mid i) = \sigma^2_i$ within the strata by prior experiments or by estimating with the respective n_i measured values: $\sigma_i^2 \approx \frac{1}{n_i - 1} \sum_{j=1}^{n_i} (x_{ij} - \bar{x}_i)^2; \; \bar{x}_i = \frac{1}{n_i} \sum_{j=1}^{n_i} x_{ij}$
5th step	Determining the sample mean from the x_{ij}:
(6.2.7)	$\bar{\bar{x}} = \frac{1}{N} \sum_{i=1}^{k} N_i \bar{x}_i = \frac{1}{N} \sum_{i=1}^{k} \left\{ N_i \cdot \frac{1}{n_i} \cdot \sum_{j=1}^{n_i} x_{ij} \right\}$
6th step	Determining the variance of the sample mean
(6.3.2a)	$\sigma_p^2 = (\frac{1}{n} - \frac{1}{N}) \cdot \frac{1}{N} \sum_{i=1}^{k} N_i \cdot \sigma_i^2$
7th step	Preassigning a confidence coefficient S(e.q. 95%)
8th step	Find the solution c from the normal distribution*):
(5.2.4)	$S = 2\,H(c) - 1$
9th step	Hence the confidence region with the confidence coefficient S and $n = \Sigma n_i$ is given by:
(5.2.6)	$\mu = x \pm c \frac{\sigma}{\sqrt{n}} \sqrt{\frac{N-n}{N-1}}$
	For $N \gg n$:
(5.2.5)	$\mu = x \pm c \frac{\sigma}{\sqrt{n}}$

*)It is assumed that the values are approximately normally distributed, otherwise equation (5.1.3) has to be used.

and hence:

$$\sigma^2 = E(\sigma_i^2) + \sigma^2(\mu_i) \qquad (6.3.4)$$

$E(\sigma_i^2)$ is the mean of all the variances of the individual strata and $\sigma^2(\mu_i)$ is the variance of the means between the strata. Equation (6.3.4) and (6.3.2) into equation (6.2.2) leads to:

$$\sigma_p^2(\bar{\bar{x}}) = \sigma^2(\bar{x}) - \left(\frac{1}{n} - \frac{1}{N}\right) \cdot \sigma^2(\mu_i) \qquad (6.3.5)$$

The variance of an unstratified sample of size n is always larger than the variance of a stratified sample of proportional subdivision having the same number of elements. The proportional subdivision is the more effective the more the means or the expectations μ_i within the various strata differ from one another.

6.4 Cost-optimizing selection for a stratified sample

Though proportional subdivision is very readily manipulated, from an economic point of view it does not always represent the optimum achievable. Two different problems have to be solved:

1) Which part sample sizes n_i minimize the variance for a given cost c?, and

2) Which part sample sizes n_i minimize cost for a given precision?

1st Problem: Given cost c (Table 6.4.1)

Though proportional subdivision produces a smaller variance than the unstratified random sample, for a given sample size n, it does not always produce the smallest possible variance. Because maximum precision is of interest in sampling investigations, the number of elements n_i of the individual strata is chosen so that equation (6.2.12) is minimized. In what follows it is always assumed that the strata are sufficiently large, i.e. $N_i \gg 1$. In order to determine an optimum, a secondary condition must be given. In general, this condition is the cost c.

If the investigation of the sample attribute of the i-th stratum costs the amount $n_i \cdot c_i$, then

$$\sum_{i=1}^{k} n_i \cdot c_i = c \qquad (6.4.1)$$

The necessary condition for a minimum of $\sigma^2(\bar{\bar{x}})$ for a given cost c is found by forming the total differential of equation (6.2.12):

$$d\,\sigma^2(\bar{\bar{x}}) = \frac{\partial\sigma^2(\bar{\bar{x}})}{\partial n_1}\,dn_1 + \frac{\partial\sigma^2(\bar{\bar{x}})}{\partial n_2}\,dn_2 + \ldots + \frac{\partial\sigma^2(\bar{\bar{x}})}{\partial n_k}\,dn_k = 0$$

(6.4.2)

Here it must be stressed that equation (6.4.2) is exact if the n_i are continuous variables. However, in fact, the n_i are discrete with a steplength of one. Therefore, equation (6.4.2) only yields exact results within this steplength. Partial differentiation results in

$$\left(\frac{N_1}{N}\right)^2 \cdot \frac{1}{n_1^2} \cdot \sigma_1^2\,dn_1 + \left(\frac{N_2}{N}\right)^2 \cdot \frac{1}{n_2^2} \cdot \sigma_2^2\,dn_2 + \ldots \left(\frac{N_k}{N}\right)^2 \cdot \frac{1}{n_k^2} \cdot \sigma_k^2\,dn_k = 0$$

(6.4.3)

Table 6.4.1 Cost-optimizing selection for a stratified sample and given cost c

1st step	Selection of strata i = 1, 2, ... k
2nd step	Fixing the size of the investigation units and hence determining the N_i and $N = \Sigma N_i$
3rd step	Fixing the costs c_i of the investigation of the sample attribute of the i-th stratum
4th step	Determining the variances $\sigma^2(x_{ij} \mid i) = \sigma^2_i$ within the strata by prior experiments or from earlier results
5th step	a) Determining the sample sizes n_i and $n = \Sigma n_i$

(6.4.6)
$$n_i = \frac{1}{\lambda} \cdot \frac{N_i}{N} \cdot \frac{\sigma_i}{\sqrt{c_i}}$$

(6.4.7)
$$\lambda = \frac{1}{c} \sum_{i=1}^{k} \frac{N_i}{N} \sigma_i \sqrt{c_i}$$

b) Determining the samples sizes n_i and $n = \Sigma n_i$ for the special case that all $c_i = c^*$

(6.4.9)
$$n_i = \frac{c}{c^*} \frac{N_i \cdot \sigma_i}{\sum_{i=1}^{k} N_i \cdot \sigma_i}$$

6th step Determining the sample mean from the x_{ij}:

(6.2.7)
$$\bar{\bar{x}} = \frac{1}{N} \sum_{i=1}^{k} N_i x_i = \frac{1}{N} \sum_{i=1}^{k} \left\{ N_i \cdot \frac{1}{n_i} \sum_{j=1}^{n_i} x_{ij} \right\}$$

7th step Determining the variance of the sample mean

(6.4.11)
$$\sigma_c^2 = \frac{1}{n} \left(\sum_{i=1}^{k} \frac{N_i}{N} \sigma_i \right)^2 - \frac{1}{N} \sum_{i=1}^{k} \frac{N_i}{N} \cdot \sigma_i^2$$

8th step Pre-assigning a confidence coefficient S(e.q. 95%)

9th step Find the solution c from the normal distribution*):

(5.2.4) $S = 2\,H(c) - 1$

10th step Hence the confidence region with the confidence coefficient S and $n = \Sigma\, n_i$ is given by:

(5.2.6) $$\mu = \bar{\bar{x}} \pm c\,\frac{\sigma}{\sqrt{n}}\sqrt{\frac{N-n}{N-1}}$$

For $N \gg n$

(5.2.5) $$\mu = \bar{\bar{x}} \pm c \cdot \frac{\sigma}{\sqrt{n}}$$

*)It is assumed that the values are approximately normally distributed, otherwise equation (5.1.3) has to be used.

If now the total differential of equation (6.4.1) is formed, then, because c is assumed to be constant:

$$dc = c_1\, dn_1 + c_2\, dn_2 + \dots + c_k\, dn_k = 0 \qquad (6.4.4)$$

If this equation is multiplied with $-\lambda^2$ and added to equation (6.4.3) then:

$$\sum_{i=1}^{k} \left(\frac{N_i}{N}\right)^2 \cdot \frac{1}{n_i^2} \cdot \sigma_i^2 \cdot dn_i = \lambda^2 \sum_{i=1}^{k} c_i\, dn_i \qquad (6.4.5)$$

Comparing the coefficients leads to

$$\left(\frac{N_i}{N}\right)^2 \cdot \frac{1}{n_i^2} \cdot \sigma_i^2 = \lambda^2 \cdot c_i$$

or in the form of a condition for n_i:

$$n_i = \frac{1}{\lambda} \cdot \frac{N_i}{N}\, \frac{\sigma_i}{\sqrt{c_i}} \qquad (6.4.6)$$

The constant λ^2 follows from condition (6.4.1):

$$\sum_{i=1}^{k} \frac{1}{\lambda} \left(\frac{N_i}{N}\right) \frac{\sigma_i}{\sqrt{c_i}} \cdot c_i = c$$

$$\boxed{\lambda = \frac{\sum_{i=1}^{k} \frac{N_i}{N} \cdot \sigma_i \cdot \sqrt{c_i}}{c}} \qquad (6.4.7)$$

It can be seen in equation (6.4.6) that the cost-optimized subdivision is achieved if the numbers n_i of the elements are proportional to the stratum proportions N_i/N and proportional to the standard deviations σ_i^2 within the strata. This relationship is readily visualized: the extent to which a stratum has to be taken into account increases with its size N_i and the inhomogeneity within the stratum, expressed by σ_i. On the other hand, increasing costs lower the number of elements, since a cost-optimized selection has to be made.

In the industrial field, the cost of investigating the samples in the individual layers are nearly always the same. In this important special case, where the costs c_i should be independent of the selection stratum, namely $c_i = c^*$, equation (6.4.1) becomes:

$$\sum_{i=1}^{k} n_i \cdot c_i = n \cdot c^* = c \qquad (6.4.8)$$

and, instead of equation (6.4.6):

$$\boxed{n_i = \frac{c}{c^*} \cdot \frac{N_i \cdot \sigma_i}{\sum_{i=1}^{k} N_i \cdot \sigma_i}} \qquad (6.4.9)$$

the variance $\sigma_k^2(\bar{\bar{x}})$ which is the smallest possible under these conditions is given by equation (6.2.12) for sufficiently large strata $N_i \gg 1$:

$$\sigma_c^2(\bar{\bar{x}}) = \sum_{i=1}^{k} \left(\frac{N_i}{N}\right)^2 \left(\frac{1}{n_i} - \frac{1}{N_i}\right) \cdot \sigma_i^2$$

$$\sigma_c^2(\bar{\bar{x}}) = \sum_{i=1}^{k} \left(\frac{N_i}{N}\right)^2 \cdot \frac{\sigma_i^2}{n_i} - \frac{1}{N} \sum_{i=1}^{k} \frac{N_i}{N} \cdot \sigma_i^2 \qquad (6.4.10)$$

With the aid of aquation (6.4.9) this leads to:

$$\sigma_c^2(\bar{\bar{x}}) = \frac{1}{n} \left(\sum_{i=1}^{k} \frac{N_i}{N} \sigma_i\right)^2 - \frac{1}{N} \sum_{i=1}^{k} \frac{N_i}{N} \cdot \sigma_i^2 \qquad (6.4.11)$$

or:

$$\boxed{\sigma_c^2(\bar{\bar{x}}) = \frac{1}{n} E^2(\sigma_i) - \frac{1}{N} E(\sigma_i^2)}$$

(Table 6.2.1, row 14)

If the sample size n is small compared with the total number of elements N, then, approximately:

$$\sigma_c^2(\bar{\bar{x}}) = \frac{1}{n} \left(\sum_{i=1}^{k} \frac{N_i}{N} \cdot \sigma_i\right)^2 = \frac{1}{n} E^2(\sigma_i) \qquad (6.4.13)$$

With the aid of the computational rule for expectations, equation (2.9.2), then:

$$E^2(\sigma_i) = E(\sigma_i^2) - \sigma^2(\sigma_i) \qquad (6.4.14)$$

$\sigma^2(\sigma_i)$ is the variance of the standard deviation σ_i which varies from stratum to stratum. It follows from equation (6.4.12) that:

$$\sigma_c^2(\bar{\bar{x}}) = \frac{1}{n} \left(E(\sigma_i^2) - \sigma^2(\sigma_i)\right) - \frac{1}{N} E(\sigma_i^2)$$

$$\sigma_c^2(\bar{\bar{x}}) = \left(\frac{1}{n} - \frac{1}{N}\right) E(\sigma_i^2) - \frac{1}{n} \cdot \sigma^2(\sigma_i) \qquad (6.4.15)$$

The first summand corresponds to the variance of the sample mean for proportional subdividion:

$$\sigma_c^2(\bar{\bar{x}}) = \sigma_p^2(\bar{\bar{x}}) - \frac{1}{n} \sigma^2(\sigma_i) \qquad (6.4.16)$$

For the same sample size n, the variance $\sigma_c^2(\bar{\bar{x}})$ for cost-optimizing subdivision and constant costs per unit of investigation is always better than in the case of proportional subdivision. If the variations σ_i within the strata are all the same, then the variance $\sigma^2(\sigma_i)$ disappears and the cost-optimized variance $\sigma_c^2(x)$ becomes the variance $\sigma_p^2(\bar{\bar{x}})$ for the case of proportional subdivision.

Despite the improvement in precision which is obtained in the case of cost-optimizing selection, the improvement (i.e. the 2nd summand in (6.4.16)) is frequently so small in practice that it must always be re-examined to determine whether a non-proportional subdivision is worth the greater expense.

2. Problem: Given precision $\sigma^2(\bar{\bar{x}}) = \sigma_v^2(\bar{\bar{x}})$ (Table 6.4.2)

The necessary condition for a cost minimum in the case of a given precision required is found by forming the total differential of equation (6.4.1):

$$dc = c_1\, dn_1 + c_2 \cdot dn_2 \ldots c_k \cdot dn_k = 0 \tag{6.4.17}$$

If the total differential of equation (6.2.12) is formed, then, because the variance $\sigma^2(\bar{\bar{x}}) = \sigma_v^2$ has been made a constant:

$$\left(\frac{N_1}{N}\right)^2 \cdot \frac{1}{n_1^2}\sigma_1^2\, dn_1 + \left(\frac{N_2}{N}\right)^2 \frac{1}{n_2^2}\sigma_2^2\, dn_2 + \ldots \left(\frac{N_k}{N}\right)^2 \frac{1}{n_k^2}\cdot\sigma_k^2\, dn_k = 0 \tag{6.4.18}$$

The same equations hence apply in the case of given precision as in the case of given costs, and the solutions are analogous. The selection condition is identical to equation (6.4.6)

$$\boxed{n_i = \frac{1}{\lambda}\,\frac{N_i}{N}\,\frac{\sigma_i}{\sqrt{c_i}}} \tag{6.4.19}$$

The constant λ is calculated, using equation (6.4.10), from

$$\sigma_v^2 = \sum_{i=1}^{k} \left(\frac{N_i}{N}\right) \cdot \sigma_i \sqrt{c_i} \cdot \lambda - \frac{1}{N} \sum_{i=1}^{k} \frac{N_i}{N} \cdot \sigma_i^2$$

$$\boxed{\lambda = \frac{\sigma_v^2 + \frac{1}{N} \sum_{i=1}^{k} \left(\frac{N_i}{N}\right) \cdot \sigma_i^2}{\sum_{i=1}^{k} \left(\frac{N_i}{N}\right) \cdot \sigma_i \sqrt{c_i}}} \qquad (6.4.20)$$

Table 6.4.2 Cost-optimizing selection for stratified sampling and given variance

1st step Selection of the strata $i = 1, 2, \ldots k$

2nd step Fixing the size of the unit of investigation and hence determining the N_i and $N = \Sigma N_i$

3rd step Fixing the costs c_i for investigating the sample attribute of the i-th stratum

4th step Determining the variances $\sigma^2(x_{ij}|i) = \sigma^2_i$ within the strata by prior experiments or from earlier results

5th step Determining the sample size n_i and $n = \Sigma n_i$

(6.4.19)
$$n_i = \frac{1}{\lambda} \frac{N_i}{N} \frac{\sigma_i}{\sqrt{c_i}}$$

(6.4.20)
$$\lambda = \frac{\sigma_v^2 + \frac{1}{N} \sum_{i=1}^{k} \left(\frac{N_i}{N}\right) \cdot \sigma_i^2}{\sum_{i=1}^{k} \left(\frac{N_i}{N}\right) \cdot \sigma_i \cdot \sqrt{c_i}}$$

6th step The minimum cost c to be expected is given by:

a)

(6.4.21)
$$c = \frac{1}{\lambda} \sum_{i=1}^{k} \left(\frac{N_i}{N}\right) \cdot \sigma_i \sqrt{c_i}$$

b) In the special case where $c_i = c^*$

(6.4.22)
$$c = n \cdot c^*$$

The minimum cost c to be expected is given by:

$$c = \frac{1}{\lambda} \sum_{i=1}^{k} \left(\frac{N_i}{N}\right) \cdot \sigma_i \cdot \sqrt{c_i} \qquad (6.4.21)$$

In the frequent special case that the costs c_i are independent of the selection stratum, namely $c_i = c^*$, the simplified equation, with $n \cdot c^* = c$, holds:

$$c = \frac{1}{n \cdot \lambda^2} \left(\sum_{i=1}^{k} \left(\frac{N_i}{N}\right) \cdot \sigma_i \right)^2 = n \cdot c^* \qquad (6.4.22)$$

6.5 Example of stratified sampling

Object: The total fat content of a block of chocolate which weighs 10 kg is to be determined. During the solidification of liquid chocolate melts, the heavier solid particles settle out, so that different layers are expected to have different fat contents. Therefore, a stratified sample appears to be advisable. Solution (Table 6.5.1).

1. Selection of the strata. The block of chocolate is divided vertically into various part blocks which weigh, for example, 1 lb, 2 lb, 3 lb and 4 lb.

2. Fixing the elements: They are usually identical to the units of investigation, for example 1 g. The population then consists of $N_1 = 1{,}000$, $N_2 = 2{,}000$, $N_3 = 3{,}000$ and $N_4 = 4{,}000$ units.

3. The variances σ_i^2 within the strata, which are required for the determination of the optimum selection procedure, must be found by prior experiments or must be known from experience.

Table 6.5.1 Worked example for a stratified sample

Strata	N_i	proportional n_i	σ_i^2	σ_i	$N_i \cdot \sigma_i^2$	$N_i \cdot \sigma_i$	$n \cdot \frac{N_i \cdot \sigma_i}{\sum N_i \sigma_i}$	n_i c = 200 $	$\frac{\sigma_i}{\sqrt{c^*}} \cdot \frac{N_i}{N}$	n_i $\sigma_c^2 \leq 10^{-4}$
1	1000	1	$10 \cdot 10^{-4}$	$3,16 \cdot 10^{-2}$	1	31,6	0,86	1	4,61	5
2	2000	2	$20 \cdot 10^{-4}$	$4,47 \cdot 10^{-2}$	4	89,4	2,44	2	13,05	13
3	3000	3	$5 \cdot 10^{-4}$	$2,24 \cdot 10^{-2}$	1,5	67,2	1,82	2	9,81	10
4	4000	4	$20 \cdot 10^{-4}$	$4,47 \cdot 10^{-2}$	8	178,8	4,87	5	26,10	26
	10000	10	$\sigma^2 = 41,4 \cdot 10^{-4}$	$\sigma = 6 \cdot 43 \cdot 10^{-2}$	14,5	367.0		10		54

a) Sampling without stratification for n = 10

Estimate $\bar{x} = \frac{1}{10} \cdot \sum_{i=1}^{10} x_i$ Variance $\sigma^2(\bar{x}) = 4,14 \cdot 10^{-4}$

b) Sampling with proportional stratification for n = 10

Estimate $\bar{\bar{x}} = \frac{1}{N} \sum_{i=1}^{k} N_i \cdot \frac{1}{n_i} \sum_{j=1}^{n_i} x_{ij}$ Variance $\sigma_p^2(\bar{\bar{x}}) = 1,45 \cdot 10^{-4}$

c) Sampling with cost-optimized stratification for c = 200 $,

c* = 20 DM and n = 10

Estimate $\bar{\bar{x}} = \frac{1}{N} \sum N_i \frac{1}{n_i} \sum x_{ij}$ Variance $\sigma_c^2(\bar{\bar{x}}) = 1,35 \cdot 10^{-4}$

d) Sampling for a given variance $\sigma_v^2(\bar{\bar{x}}) \leq 2,5 \cdot 10^{-5}$, c* = 20 $

$\sqrt{c^*} \cdot \lambda = 6,85 \cdot 10^{-4}$; $c = \sum_{i=1}^{k} n_i c^* = 1080$ $; $\sigma^2(\bar{\bar{x}}) = 2,5 \cdot 10^{-5}$

Let $\sigma_1^2 = 10 \times 10^{-4}$, $\sigma_2^2 = 20 \times 10^{-4}$, $\sigma_3^2 = 5 \times 10^{-4}$ and $\sigma_4^2 = 20 \times 10^{-4}$. Hence the variance of the population is given by $\sigma^2 = 41.4 \times 10^{-4}$.

4. Choice of the selection procedure:
a) Without stratification, for example for n = 10 elements, each of 1 g: Estimate for the population:

$$\bar{x} = \frac{1}{10} \sum_{l=1}^{10} x_l$$

Equation (5.2.5), with a confidence coefficient of 95 % and $\sigma^2 = 41.4 \times 10^{-4}$, leads to:

$$\mu = x \pm 0.04 \qquad (6.5.1)$$

b) Sampling with proportional subdivision for n = 10. The estimate for the expectation μ of the population is given by:

$$\bar{\bar{x}} = \frac{1}{N} \sum_{i=1}^{k} N_i \cdot \frac{1}{n_i} \sum_{j=1}^{n_i} x_{ij}$$

The subdivision is proportional, by equation (6.3.1):

$n_1 = 1$, $n_2 = 2$, $n_3 = 3$, $n_4 = 4$

Hence the variance is given by equation (6.3.2a) as:

$$\sigma^2_p = 1.45 \times 10^{-4}$$

and then, with a confidence coefficient of 95 %:

$$\mu = \bar{\bar{x}} \pm 0.02 \qquad (6.5.2)$$

c) Sampling with optimum subdivision for given costs of c = 200 \$ and analysis costs per unit of c* = 20 \$. Hence, by equation (6.4.9):

$n_1 = 0.86$, $n_2 = 2.44$, $n_3 = 1.83$, $n_4 = 4.87$.

In deriving the formulae, the numbers n_i were assumed to be continuous. Therefore, non-integral values for the n_i are therefore also possible. In practice the following numbers are thus chosen: $n_1 = 1$, $n_2 = 2$, $n_3 = 2$, $n_4 = 5$ and $\sum_{i=1}^{4} n_i = 10$. Hence equation (6.3.2a) becomes:

$$\sigma_c^2 = 1.35 \times 10^{-4}$$

In deriving this equation the n_i were not yet assumed to be continuous; equation (6.3.2a) therefore holds exactly for the chosen numbers. A confidence coefficient of 95 % then gives:

$$\mu = \bar{\bar{x}} \pm 0.02 \qquad (6.5.3)$$

The cost-optimized selection does not represent a (substantial) improvement compared with the proportional selection, in the example chosen here.

d) If an interval is given for this problem, with a confidence coefficient of 95 %, for example:

$$\bar{\bar{x}} - 0.01 \leqq \mu \leqq \bar{\bar{x}} + 0.01$$

this requires the variance of

$$\sigma_v^2(\bar{\bar{x}}) \leqq 2.5 \times 10^{-5}$$

With the same costs, namely $c_i = c^* = 20$ \$, equation (6.4.20) gives:

$$\lambda \cdot \sqrt{c^*/\$} = 6.85 \times 10^{-4}$$

and hence from equation (6.4.19):

$$n_1 = 4.61,\ n_2 = 13.05,\ n_3 = 9.81,\ n_4 = 26.10.$$

The statement already made in c) also holds for the n_i. If, analogously, $n_1 = 5$, $n_2 = 13$, $n_3 = 10$ and $n_4 = 26$ are chosen,

then:

$$\sigma^2(\bar{\bar{x}}) = 2.5 \times 10^{-5}.$$

Rounding-off to give discrete sample numbers n_i may, in an unfavorable case (in particular for small sample sizes n), produce variances which are a little larger. This must be taken into account in each individual case. Roundig-up all the numbers is, however, always safe. However, this increases the cost of the sample in this example from

$$c_1 = \sum_{i=1}^{4} n_i \cdot c^* = 1{,}080 \ \$$$

in the first case to

$$c_2 = 1{,}120 \ \$$$

for the sample numbers of $n_1 = 5$, $n_2 = 14$, $n_3 = 10$ and $n_4 = 27$. On the other hand, the variance is

$$\sigma_v^2 = 2.4 \times 10^{-5}$$

and is thus certainly on the safe side.

6.6 Multi-stage sampling (Tables 6.6.1 and 6.6.2)

Bulk goods, such as coal, ore and the like, are characterized according to their attributes, such as ash content, water content, metal content and the like. This requires that analytical samples which usually only weigh a few grams have to be obtained from the usually large bulk amounts, for example 1,000 t, in such a way that they permit inferences about the characteristics of the bulk material. A simple random sample of the order of magnitude of an analytical sample is usually not possible here, if only because only fine material could possibly be taken into account because of its small size. The mass of the individual samples to be taken should be at least large enough to make it possible that several

Table 6.6.1 Multi-stage sampling

		Population	Total sample	Actual sample
1st stage	Number of elements	N_1	n_1	n_1
	Attribute of the elements	$1 \leq i \leq N_1$ μ_i	$1 \leq i \leq n_1$ μ_i	$1 \leq i \leq n_1$ x_i
	Expectation or mean	$E(\mu_i) = E(x_i) = \mu$	$\bar{x}^* = \frac{1}{n_1}\sum_{i=1}^{n_1} \mu_i$	$\bar{x} = \frac{1}{n_1}\sum_{i=1}^{n_1} x_i$
	Variance	$\sigma^2(\mu_i) = \sigma_1^2$	$\sigma^2(\bar{x}^*) = \frac{N_1 - n_1}{N_1 - 1} \cdot \frac{\sigma_1^2}{n_1}$	
2nd stage	Number of elements	N_2	n_2	n_2
	Serial numbers of the elements	$1 \leq j \leq N_2$	$i \leq j \leq n_2$	$1 \leq j \leq n_2$
	Attribute of the elements	μ_{ij}	μ_{ij}	x_{ij}
	Expectation or mean	$E(\mu_{ij}\|i) = E(x_{ij}\|i) = \mu_i$	$x_i^* = \frac{1}{n_2}\sum_{j=1}^{n_2} \mu_{ij}$	$x_i = \frac{1}{n_2}\sum_{j=1}^{n_2} x_{ij}$
	Variance	$\sigma^2(\mu_{ij}\|i) = \sigma_2^2$	$\sigma^2(x_i^*\|i) = \frac{N_2 - n_2}{N_2 - 1} \cdot \frac{\sigma_2^2}{n_2}$	
3rd stage	Number of elements	N_3	n_3	n_3
	Serial numbers of the elements	$1 \leq k \leq N_3$	$1 \leq k \leq n_3$	$1 \leq k \leq n_3$
	Attribute of the elements	μ_{ijk}	μ_{ijk}	x_{ijk}
	Expectation or mean	$E(\mu_{ij}\|i\|j) = \mu_{ij}$	$x_{ij}^* = \frac{1}{n_3}\sum_{k=1}^{n_3} \mu_{ijk}$	$x_{ij} = \frac{1}{n_3}\sum_{k=1}^{n_3} x_{ijk}$
	Variance	$\sigma^2(\mu_{ijk}\|i\|j) = \sigma^2_3$	$\sigma^2(x_{ij}^*\|i\|j) = \frac{N_3 - n_3}{N_3 - 1} \cdot \frac{\sigma_3^2}{n_3}$	

Table 6.6.2 Multi-stage sampling

1st step	Fixing the stage division 1, 2, 3
2nd step	Fixing the size of the units of investigation in each stage and hence determining the N_1, N_2, N_3 and $N = N_1 \cdot N_2 \cdot N_3$
3rd step	Selecting the sample sizes n_1, n_2 and n_3 in each stage and $n = n_1 \cdot n_2 \cdot n_3$
4th step	Determining the sample mean

$$\bar{x} = \frac{1}{n_1 n_2 n_3} \sum_{i=1}^{n_1} \sum_{j=1}^{n_2} \sum_{k=1}^{n_3} x_{ijk} \qquad (6.6.1)$$

5th step — Determining the variances σ_1^2, σ_2^2 and σ_3^2 in the stages by prior experiments or by estimation based on the respective measured values:

$$\sigma_3^2 \approx s_3^2 = \frac{1}{n_1 n_2 (n_3-1)} \sum_{i=1}^{n_1} \sum_{j=1}^{n_2} \sum_{k=1}^{n_3} (x_{ijk} - x_{ij})^2 ; \; x_{ij} = \frac{1}{n_3} \sum_{k=1}^{n_3} x_{ijk}$$

$$\sigma_2^2 \approx s_2^2 = \frac{1}{n_1 (n_2-1)} \sum_{i=1}^{n_2} (x_{ij} - x_i)^2 - \frac{s_3^2}{n_3} ; \; x_i = \frac{1}{n_2} \sum_{j=1}^{n_2} x_{ij}$$

$$\sigma_1^2 \approx s_1^2 = \frac{1}{n_1 - 1} \sum_{i=1}^{n_1} (x_i - \bar{x}) - \frac{s_2^2}{n_2} - \frac{s_3^2}{n_2 n_3} ; \; \bar{x} = \frac{1}{n_1} \sum_{i=1}^{n_1} x_i$$

6th step — Determining the variance of the sample mean:

a)

$$\sigma^2(\bar{x}) = \frac{N_1 - n_1}{N_1 - 1} \frac{\sigma_1^2}{n_1} + \frac{1}{n_1} \frac{N_2 - n_2}{N_2 - 1} \frac{\sigma_2^2}{n_2} + \frac{1}{n_1 n_2} \frac{N_3 - n_3}{N_3 - 1} \frac{\sigma_3^2}{n_3} \qquad (6.6.23)$$

b) for $N_1 \gg n_1$

$$\sigma^2(\bar{x}) = \frac{\sigma_1^2}{n_1} + \frac{\sigma_2^2}{n_1 \cdot n_2} + \frac{\sigma_3^2}{n_1 n_2 n_3} \qquad (6.6.24)$$

7th step — Pre-assigning a confidence coefficient S(e.q. 95%).

8th step Find the solution c from the normal distribution*):

(5.2.4) $$S = 2\,H(c) - 1$$

9th step Hence the confidence region with the confidence coefficient S and $n = \Sigma n_i$ is given by:

(5.2.6) $$\mu = \bar{x} \pm c \frac{\sigma}{\sqrt{n}} \sqrt{\frac{N-n}{N-1}}$$

For $N \gg n$:

(5.2.5) $$\mu = \bar{x} \pm c \frac{\sigma}{\sqrt{n}}$$

*) It is assumed that the values are approximately normally distributed, otherwise equation 5.1.3 must be used.

pieces of the largest particle size manage to get into the individual sample (see section 7.3) (sheet 2 of DIN 51,061, specifies that the removed sample must have a weight which is at least ten times that of the largest particle).

In practice, a population frequently consists of certain units, for example bags, barrels or boxes. If from these N_1 "units of the 1st stage" a subset of n_1 such elements is taken as a sample, then these elements are however much too large for analysis.

Therefore, each of the units of the 1st stage chosen for the sample is divided, in each case, into N_2 units of the 2nd stage, of which n_2 elements are, in each case, selected to form a sample (Figure 6.6.1). If these units are still too large, then this procedure can be continued until a size of sample appropriate for analysis has been reached. How the individual stages are subdivided may be predetermined by technical factors, for example

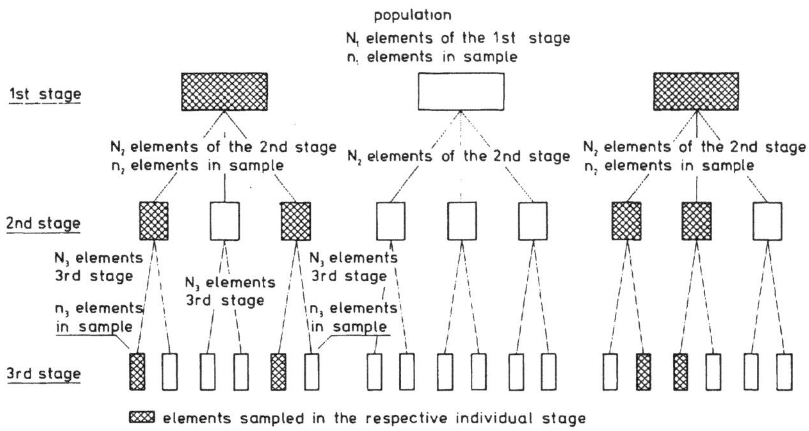

Figure 6.6.1 Multi-stage sampling scheme

by packaged units, but subdivision may also be effected in the case of lumpy material after comminution and mixing in a sample divider. The whole procedure from the elements of the first stage to the elements of the analytical samples is referred to as sample preparation and sample division. In contrast to the population statistics and economic statistics, where the units of the first or also of the further stages usually have very different sizes since they are natural units, units in industry are usually of the same or nearly the same size. A further frequently possible simplification is that the variances within each subset. The subsets being chosen so that they are of the same size, are the same. Whereas the first condition can be influenced by the operator, the second condition has to be checked in cases of doubt. Inferences about the population must be based on analytical samples which, in the end, are frequently small. Here, just as in the case of the stratified sample, the question concerning the degree of confidence and the optimum choice of sample sizes arises. The terms which are required for the solution have been collated in Table 6.6.1.

Only the samples of the last unit are analyzed. The mean of these samples serves as an estimate for the population. For example, a three-stage sample leads to

$$\bar{x} = \frac{1}{n_1 \cdot n_2 \cdot n_3} \sum_{i=1}^{n_1} \sum_{j=1}^{n_2} \sum_{k=1}^{n_3} x_{ijk} \qquad (6.6.1)$$

In the case of simple random sampling from $n = n_1 \cdot n_2 \cdot n_3$ samples of the size of the units of the final stage

$$\sigma^2(\bar{x}) = \frac{N-n}{N-1} \cdot \frac{\sigma^2}{n} \qquad (6.6.2)$$

holds. Here, σ^2 is the variance of the populaton and N the total number of the elements of the final stage which are contained in the population. In multi-stage sampling, the elements of each stage are no longer selected independently. The variance of the sample mean of a multi-stage sample, which is formed according to equation (6.6.1), does not therefore agree with the variance of simple random sampling according to equation (6.6.2). The object of multi-stage sampling is hence to make a selection which permits a confidence level which is as high as possible. The influence which the sample size of each stage has on the variance $\sigma^2(x)$ to be expected, and how many units of each stage have to be used in the investigations in order, for example, to achieve a cost-optimized result must be examined. It follows directly from what was demonstrated above that the expectation $\bar{\bar{x}}$ corresponds to the expectation μ of the population.

$$E(\bar{x}) = \mu \qquad (6.6.3)$$

The first selection stage consists of N_1 elements of the 1st stage with the attributes μ_i and the variance $\sigma^2(\mu_i) = \sigma_1^2$. If n_1 of these elements were taken as a sample, then the mean of this total sample would be

$$\bar{x}_1^* = \frac{1}{n_1} \sum_{i=1}^{n_1} \mu_i \qquad (6.6.4)$$

and the variance, which corresponds to sampling without replacement, is given by:

$$\sigma^2(\bar{x}^*) = \frac{N_1 - n_1}{N_1 - 1} \cdot \frac{\sigma_1^2}{n_1} \tag{6.6.5}$$

In contrast to this total sample, only subsets, namely, in each case, N_2 elements of the second stage, are taken when actually sampling the N_1 elements. Here the values μ_i are estimated by the values x_i and the mean of the actual sample, which is formed from the latter, is given by

$$\bar{x} = \frac{1}{n_1} \sum_{i=1}^{n_1} x_i \tag{6.6.6}$$

The x_i are means of the 2nd stage:

$$x_i = \frac{1}{n_2} \sum_{j=1}^{n_2} x_{ij} \tag{6.6.7}$$

etc.

The sample mean $\bar{x}$ deviates from the mean $\bar{x}^*$ of a total sample of the 1st step:

$$\bar{x} = \bar{x}^* + (\bar{x} - \bar{x}^*) \tag{6.6.8}$$

If the quantities are independent then:

$$\sigma^2(\bar{x}) = \sigma^2(\bar{x}^*) + \sigma^2(\bar{x} - \bar{x}^*) \tag{6.6.9}$$

The 1st term, $\sigma^2(x^*)$, on the right-hand side is given by equation (6.6.5). In order to calculate the 2nd term equations (6.6.4) and (6.6.6) are used:

$$\sigma^2(\bar{x} - \bar{x}^*) = \sigma^2 \left(\frac{1}{n_1} \sum_{i=1}^{n_1} (x_i - \mu_i)\right)$$

$$\sigma^2(\bar{x} - \bar{x}^*) = \frac{1}{n_1} \cdot \sigma^2(x_i - \mu_i) \tag{6.6.10}$$

$\sigma^2(x_i - \mu_i)$ is the variance of the deviation of the sample values of the 1st stage. Since this variance is assumed to be the same for each element of the 1st stage and to be independet of i:

$$\sigma^2(x_i - \mu_i) = \sigma^2(x_i - \mu_i | i) \tag{6.6.11}$$

but, with the condition i, μ_i is constant and hence

$$\sigma^2(x_i - \mu_i | i) = \sigma^2(x_i | i) \tag{6.6.12}$$

x_i are the means of the values of the 2nd stage, which in turn are formed as means of some values of the 3rd stage. They too deviate from the mean x_i^* of all values of the 2nd stage:

$$x_i = x_i^* + (x_i - x_i^*) \tag{6.6.13}$$

Again, independence of the quantities is assumed:

$$\sigma^2(x_i | i) = \sigma^2(x_i^* | i) + \sigma^2(x_i - x_i^* | i) \tag{6.6.14}$$

The 1st term on the right-hand side $\sigma^2(x_i^* | i)$ corresponds to the variance of n_2 exact values of the elements of stage 2 and:

$$\sigma^2(x_i^* | i) = \frac{N_2 - n_2}{N_2 - 1} \cdot \frac{\sigma_2^2}{n_2} \tag{6.6.15}$$

The calculation of the 2nd term proceeds analogously to the 1st stage:

$$\sigma^2(x_i - x_i^* | i) = \sigma^2 \left(\frac{1}{n_2} \sum_{j=1}^{n_2} (x_{ij} - \mu_{ij}) \middle| i \right)$$

$$\sigma^2(x_i - x_i^* | i) = \frac{1}{n_2} \cdot \sigma^2(x_{ij} - \mu_{ij} | i) \qquad (6.6.16)$$

$\sigma^2(x_{ij} - \mu_{ij} | i)$ is the vaiance of the deviation of the sample values of the 2nd stage. Again, since this variance is assumed to be the same for each element of the 2nd stage and to be independent of j:

$$\sigma^2(x_{ij} - \mu_{ij} | i) = \sigma^2(x_{ij} - \mu_{ij} | i | j) \qquad (6.6.17)$$

Hence the μ_{ij} on the right-hand side are constant:

$$\sigma^2(x_{ij} - \mu_{ij} | i | j) = \sigma^2 (x_{ij} | i | j) \qquad (6.6.18)$$

x_{ij} are the means from the values of the 3rd stage, which in turn can be the means of some values of the 4th stage. They then deviate from the mean x_{ij} of all the values of the 3rd stage:

$$x_{ij} = x_{ij}^* + (x_{ij} - x_{ij}^*) \qquad (6.6.19)$$

The independence of the quantities leads to:

$$\sigma^2(x_{ij} | i | j) = \sigma^2(x_{ij}^* | i | j) + \sigma^2(x_{ij} - x_{ij}^* | i | j) \qquad (6.6.20)$$

The 1st term on the right-hand side $\sigma^2(x_{ij}^* \; i | j)$ corresponds to the variance from n_3 exact values of the elements of stage 3 and:

$$\sigma^2(x_{ij}^* | i | j) = \frac{N_3 - n_3}{N_3 - 1} \cdot \frac{\sigma_3^2}{n_3} \qquad (6.6.21)$$

This procedure is continued until the final stage. The attributes of the sample then agree with the exact values. If, for example, the third stage is the final stage, then:

$$x_{ij} = x_{ij}^* \qquad (6.6.22)$$

Successive substitutions in equations (6.6.9) - (6.6.21) lead to:

$$\sigma^2(\bar{x}) = \frac{N_1-n_1}{N_1-1} \cdot \frac{\sigma_1^2}{n_1} + \frac{1}{n_1} \frac{N_2-n_2}{N_2-1} \cdot \frac{\sigma_2^2}{n_2} + \frac{1}{n_1\, n_2} \frac{N_3-n_3}{N_3-n_1} \cdot \frac{\sigma_3^2}{n_3} \qquad (6.6.23)$$

If the number of elements N_i in each stage is large compared to the sample sizes n_i, then the result reduces to:

$$\sigma^2(\bar{x}) = \frac{\sigma_1^2}{n_1} + \frac{\sigma_2^2}{n_1 \cdot n_2} + \frac{\sigma_3^2}{n_1 \cdot n_2 \cdot n_3} \qquad (6.6.24)$$

6.7 Cost-optimizing selection in multi-stage sampling (Table 6.7.1)

Two different tasks can also be set here, as was the case in arranging the sample plan for stratified sampling. On the one hand, the resulting precision should be as high as possible for a given cost. On the other hand, the cost should be as low as possible for a given precision. It is always assumed below that the stages are large compared with the part samples, namely that $N_i \gg n_i$.

1st problem: Given cost c

Table 6.7.1 Cost-optimizing selection for multi-stage sampling with given cost c

1st step	Fixing the stage subdivision 1, 2, 3
2nd step	Fixing the size of the units of investigation in each stage and hence determining the N_1, N_2, N_3 and $N = N_1 \cdot N_2 \cdot N_3$
3rd step	Fixing the costs c_i of processing an element of the i-th stage
4th step	Determining the variances σ_1^2, σ_2^2, σ_3^2 and in the stages by prior experiments (see Table 6.6.2)
5th step	Determining the sample sizes n_1, n_2, n_3 and $n = n_1 \cdot n_2 \cdot n_3$
(6.7.9)	$n_1 = \frac{c}{\sqrt{c_1}\cdot\sigma_1 + \sqrt{c_2}\,\sigma_2 + \sqrt{c_3}\,\sigma_3}\ \frac{\sigma_1}{\sqrt{c_1}}$
(6.7.7)	$n_2 = \sqrt{\frac{c_1}{c_2}}\cdot\frac{\sigma_2}{\sigma_1}$ $n_3 = \sqrt{\frac{c_2}{c_3}}\cdot\frac{\sigma_3}{\sigma_2}$
6th step	Determining the sample mean $\bar{x}$
(6.6.1)	$\bar{x} = \frac{1}{n_1 n_2 n_3} \cdot \sum_{i=1}^{n_1}\sum_{j=1}^{n_2}\sum_{k=1}^{n_3} x_{ijk}$
7th step	Determining the variance of the sample mean
(6.7.10)	$\sigma^2(\bar{x}) = \left(\sqrt{\frac{c_1}{c}}\cdot\sigma_1 + \sqrt{\frac{c_2}{c}}\cdot\sigma_2 + \sqrt{\frac{c_3}{c}}\cdot\sigma_3\right)^2$
8th step	Pre-assigning a confidence coefficient S(e.q. 95%)
9th step	Find the solution c′ from the normal distribution*):
(5.2.4)	$S = 2\,H(c') - 1$

*) It is assumed that the values are approximately normally distributed, otherwise equation (5.1.3) has to be used.

10th step Hence the confidence region with confidence coefficient S and $n = \sum_{i=1}^{k} n_i$ is given by:

(5.2.6)
$$\mu = x \pm c' \frac{\sigma}{\sqrt{n}} \sqrt{\frac{N-n}{N-1}}$$

For $N \gg n$:

(5.2.5)
$$\mu = x \pm c' \frac{\sigma}{\sqrt{n}}$$

If the costs of processing an element of the i-th stage are c_i, then the total cost for the example of a three-stage sample is:

$$c = n_1 \cdot c_1 + n_1 \cdot n_2\, c_2 + n_1 \cdot n_2 \cdot n_3 \cdot c_3 \qquad (6.7.1)$$

The number k_i of the elements of the i-th stage is to be chosen in such a way that $\sigma^2(\bar{x})$ is minimized. The necessary condition for a minimum when a fixed cost is given is obtained by forming the total differential of equation (6.6.24):

$$d\sigma^2(\bar{x}) = \frac{\partial \sigma^2(\bar{x})}{\partial n_1}\, dn_1 + \frac{\partial \sigma^2(\bar{x})}{\partial n_2}\, dn_2 + \frac{\partial \sigma^2(\bar{x})}{\partial n_3}\, dn_3 = 0 \qquad (6.7.2)$$

The n_i are treated here as continuous quantities. In fact, the n_i are discrete and have a step length of one. Equation (6.7.2) therefore yields exact results only within this step length. Partial differentiation gives:

$$d\sigma^2(\bar{x}) = -\frac{1}{n_1}\left(\frac{\sigma_1^2}{n_1} + \frac{\sigma_2^2}{n_1 n_2} + \frac{\sigma_3^2}{n_1 n_2 n_3}\right) dn_1 - \frac{1}{n_2}\left(\frac{\sigma_2^2}{n_1 n_2} + \frac{\sigma_3^2}{n_1 n_2 n_3}\right) dn_2$$

$$- \frac{1}{n_3}\; \frac{\sigma_3^2}{n_1 n_2 n_3}\, dn_3 = 0 \qquad (6.7.3)$$

If the total differential of equation (6.7.1) is formed, then, because c is constant:

$$dc = \frac{1}{n_1}(n_1 c_1 + n_1 n_2 \cdot c_2 + n_1 n_2 n_3 \cdot c_3)\, dn_1 + \frac{1}{n_2}(n_1 n_2\, c_2 + n_1 n_2 n_3 c_3)\, dn_2$$

$$+ \frac{1}{n_3}\, n_1 n_2 n_3\, c_3\, dn_3 = 0 \qquad (6.7.4)$$

If this equation is multiplied by $+\lambda^2$ and the product is added to equation (6.7.3), then, by comparing the coefficients:

$$\text{a)} \quad \lambda^2 (n_1 c_1 + n_1 n_2 \cdot c_2 + n_1 n_2\, n_3 \cdot c_3) = \left(\frac{\sigma_1^2}{n_1} + \frac{\sigma_2^2}{n_1 n_2} + \frac{\sigma_3^2}{n_1 n_2 n_3}\right)$$

$$\text{b)} \quad \lambda^2 (n_1 n_2 c_2 + n_1 n_2 n_3 c_3) = \left(\frac{\sigma_2^2}{n_1 n_2} + \frac{\sigma_3^2}{n_1 n_2 n_3}\right) \qquad (6.7.5)$$

$$\text{c)} \quad \lambda^2\, n_1 n_2 n_3 c_3 = \frac{\sigma_3^2}{n_1 n_2 n_3}$$

These equations lead to:

$$\text{a)} \quad \lambda \cdot n_1 = \frac{\sigma_1}{\sqrt{c_1}}$$

$$\text{b)} \quad \lambda \cdot n_1 \cdot n_2 = \frac{\sigma_2}{\sqrt{c_2}} \qquad (6.7.6)$$

$$\text{c)} \quad \lambda\, n_1 \cdot n_2 \cdot n_3 = \frac{\sigma_3}{\sqrt{c_3}}$$

Hence n_2 and n_3 can be found directly:

$$\text{a)} \quad n_2 = \sqrt{\frac{c_1}{c_2}} \, \frac{\sigma_2}{\sigma_1}$$

$$\text{b)} \quad n_3 = \sqrt{\frac{c_2}{c_3}} \, \frac{\sigma_3}{\sigma_2} \qquad (6.7.7)$$

It is interesting that n_2 and n_3 depend only on the ratio of the costs of the stages but not on the total cost. The larger the standard deviation within the stage, the more elements of this stage have to be selected.

In order to calculate n_1, the factor λ must be determined. For this purpose the equations 6.7.6 are substituted into the total costs (6.7.1) and hence:

$$\sqrt{c_1} \cdot \sigma_1 + \sqrt{c_2} \cdot \sigma_2 + \sqrt{c_3} \cdot \sigma_3 = \lambda \cdot c \qquad (6.7.8)$$

Hence n_1 is

$$n_1 = \frac{c}{\sqrt{c_1} \cdot \sigma_1 + \sqrt{c_2} \cdot \sigma_2 + \sqrt{c_3} \cdot \sigma_3} \cdot \frac{\sigma_1}{\sqrt{c_1}} \qquad (6.7.9)$$

The minimum variance to be expected for a given cost is therefore:

$$\sigma^2(\bar{x}) = \left(\sqrt{\frac{c_1}{c}} \cdot \sigma_1 + \sqrt{\frac{c_2}{c}} \cdot \sigma_2 + \sqrt{\frac{c_3}{c}} \cdot \sigma_3\right)^2 \qquad (6.7.10)$$

Frequently, the expense of dividing or blending the sample increases from stage to stage, so that the cost per element increases from stage to stage:

$$c_1 \leq c_2 \leq c_3 \tag{6.7.11}$$

At the same time, the homogeneity of the elements of higher stages usually improves steadily and therefore:

$$\sigma_1^2 \geq \sigma_2^2 \geq \sigma_3^2 \tag{6.7.12}$$

For this case, which frequently occurs in industry, the sample sizes are determined by equation (6.7.7):

$$\text{a)} \quad n_2 \leq 1 \qquad \text{b)} \quad n_3 \leq 1 \tag{6.7.13}$$

Since here the discrete nature of the values has to be taken into account, n_2 and n_3 cannot be smaller than 1:

$$\boxed{n_2 = n_3 = 1} \tag{6.7.14}$$

This means that, in the case presented, only one element is selected form the elements of a particular stage, for further investigations. The discrete values n_2 and n_3 cannot reach the smallest variance which is theoretically possible for continuous variables. Rather, by equation (6.7.1):

$$c = (c_1 + c_2 + c_3) \cdot n_1 \tag{6.7.15}$$

and hence

$$\boxed{n_1 = \frac{c}{c_1 + c_2 + c_3}} \tag{6.7.16}$$

Therefore, the variance is given by:

$$\sigma^2(\bar{x}) = \frac{1}{n_1}(\sigma^2_1 + \sigma^2_2 + \sigma^2_3) = \frac{c_1+c_2+c_3}{c}(\sigma_1^2 + \sigma_2^2 + \sigma_3^2) \qquad (6.7.17)$$

<u>2nd problem: Given precision $\sigma^2(\bar{x}) = \sigma_v^2$</u>

The necessary condition for minimizing the costs for a given precision is found by forming the total differential of equation (6.7.1)

$$dc = \frac{1}{n_1}(n_1 c_1 + n_1 n_2 c_2 + n_1 n_2 n_3 c_3) \cdot dn_1 + \frac{1}{n_2}(n_1 n_2 c_2 + n_1 n_2 n_3 c_3) \cdot dn_2 + \frac{1}{n_3}(n_1 n_2 n_3 c_3) \cdot dn_3 = 0 \qquad (6.7.18)$$

The total differential of equation (6.6.24) also disappears because it is assumed that the variance $\sigma^2(x)$ is constant:

$$d\sigma^2_v(\bar{x}) = -\frac{1}{n_1}\left(\frac{\sigma_1^2}{n_1} + \frac{\sigma_2^2}{n_1 n_2} + \frac{\sigma_3^2}{n_1 n_2 n_3}\right) \cdot dn_1 - \frac{1}{n_2}\left(\frac{\sigma_2^2}{n_1 n_2} + \frac{\sigma_3^2}{n_1 n_2 n_3}\right) \cdot dn_2 - \frac{1}{n_3}\frac{\sigma_3^2}{n_1 n_2 n_3} \cdot dn_3 = 0 \qquad (6.7.19)$$

Hence, for the case where the precision is given, the same equations hold as for the case where the cost is given, and the solutions are analogous. The selection instructions are identical to equations (6.7.6) and (6.7.7). In order to calculate n_1 the factor λ must again be determined. By substituting (6.7.6) into (6.6.24) it follows that:

$$\lambda = \frac{\sigma_v^2}{\sigma_1 \sqrt{c_1} + \sigma_2 \sqrt{c_2} + \sigma_3 \sqrt{c_3}} \qquad (6.7.20)$$

Table 6.7.2 Cost-optimizing selection for a multi-stage sample with given precision σ_v^2

1st step	Fixing the stage subdivision 1, 2, 3
2nd step	Fixing the size of the units of investigation in each stage and hence determining the N_1, N_2, N_3 and $N = N_1 \cdot N_2 \cdot N_3$
3rd step	Fixing the cost c_i of processing an element of the i-th stage
4th step	Determining the variances σ_1^2, σ_2^2, σ_3^2 and in stages by prior experiments (see Table 6.6.2)
5th step	Determining the sample sizes n_1, n_2, n_3 and $n = n_1 \cdot n_2 \cdot n_3$

(6.7.21)

$$n_1 = \frac{\sigma_1}{\sqrt{c_1}} \quad \frac{(\sigma_1 \sqrt{c_1} + \sigma_2 \sqrt{c_2} + \sigma_3 \sqrt{c_3})}{\sigma_v^2}$$

$$n_2 = \sqrt{\frac{c_1}{c_2}} \cdot \frac{\sigma_2}{\sigma_1}$$

$$n_3 = \sqrt{\frac{c_2}{c_3}} \cdot \frac{\sigma_3}{\sigma_2}$$

6th step The minimum cost c which is to be expected is given by:

(6.7.23)

$$c = \frac{(\sqrt{c_1} \cdot \sigma_1 + \sqrt{c_2} \cdot \sigma_2 + \sqrt{c_3} \cdot \sigma_3)}{\sigma_v^2}$$

Hence, the multi-stage sample is initiated with the n_1 being given by:

$$n_1 = \frac{\sigma_1}{\sqrt{c_1}} \cdot \frac{(\sigma_1\sqrt{c_1} + \sigma_2\sqrt{c_2} + \sigma_3\sqrt{c_3})}{\sigma_v^2} \quad (6.7.21)$$

and the minimum cost is calculated to be:

$$c = \frac{(\sqrt{c_1}\cdot\sigma_1 + \sqrt{c_2}\cdot\sigma_2 + \sqrt{c_3}\cdot\sigma_3)^2}{\sigma_v^2} \quad (6.7.22)$$

For the case where higher order staged only contribute 1 element each to the investigation ($n_2 = n_3 = 1$), analogously to equation (6.7.12):

$$c = (c_1 + c_2 + c_3) \cdot \frac{\sigma_1^2 + \sigma_2^2 + \sigma_3^2}{\sigma_v^2} \quad (6.7.23)$$

6.8 Example of multi-stage sampling

Stange[9)] gives an example where the ash content of G = 1,000 t of raw coal was to be determined. The quantity delivered consisted of N_1 = 20 cars, each containing G_1 = 50 t. From these 20 cars, n_1 = 10 cars were selected for sampling. During loading via a conveyor belt, three samples of 1.6 kg each were taken from each of these 10 cars. Accordingly, a carload consisted of N_2 = 31,000 units of 1.6 kg each. These samples were comminuted in such a manner that the ash content was not changed and were then divided, in a sample divider, into 16 samples of 100 g each, of which 2 were selected as a sample. The 100 g samples in turn were further comminuted and divided into 100 analytical samples of 1 g each and, for each 100 g sample, 2,1 g samples were ashed in a muffle furnace.

The subdivision was as follows:

1. stage	$N_1 = 20$	$n_1 = 10$	$\sigma_1^2 \approx 20 \cdot 10^{-4}$
2. stage	$N_2 = 31\,000$	$n_2 = 3$	$\sigma_2^2 \approx 5 \cdot 10^{-4}$
3. stage	$N_3 = 16$	$n_3 = 2$	$\sigma_3^2 \approx 0{,}09 \cdot 10^{-4}$
4. stage	$N_4 = 100$	$n_4 = 2$	$\sigma_4^2 \approx 0{,}0016 \cdot 10^{-4}$

The σ_i^2 were estimated by the sample variances s_i^2 of the ash content x (see Table 6.6.2).

Except for the 1st step, the finite population correction factor can be ignored and, by equation (6.6.23) or (6.6.24):

$$\sigma^2(\bar{x}) = \frac{N_1 - n_1}{N_1 - 1} \cdot \frac{\sigma_1^2}{n_1} + \frac{\sigma_2^2}{n_1 \cdot n_2} + \frac{\sigma_3^2}{n_1 n_2 n_3} + \frac{\sigma_4^2}{n_1 n_2 n_3 n_4} \qquad (6.8.1)$$

Hence:

$$\sigma^2(\bar{x}) = \frac{10}{19} \cdot \frac{20 \cdot 10^{-4}}{10} + \frac{5 \cdot 10^{-4}}{10 \cdot 3} + \frac{0{,}09 \cdot 10^{-4}}{10 \cdot 3 \cdot 2} + \frac{0{,}0016 \cdot 10^{-4}}{10 \cdot 3 \cdot 2 \cdot 2}$$

$$\sigma^2(\bar{x}) = 1{,}22 \cdot 10^{-4} \qquad (6.8.2)$$

The mean ash content $\bar{x}$ of the $n = n_1 \cdot n_2 \cdot n_3 \cdot n_4 = 120$ analyzed samples has, with a confidence coefficient of 95 %, a confidence interval of

$$\mu = \bar{x} \pm 2{,}2 \cdot 10^{-2} \qquad (6.8.3)$$

Whether the division used in this example was reasonable will be checked by estimating costs:

The division of the population into units of the 1st stage is given by the mode of transport and does not require any further expense for the sampling. The cost of the selection is negligible because of the way loading is carried out with the conveyor belt, that is to say $c_1 = 0$.

Comminution, which is often expensive, and sample division create costs per unit of processing the units of the 2nd and 3rd stages which are often considerably higher than the costs of analysis of the units of the 4th stage, namely $c_4 = c^*$. It is assumed that $c_2 = c_3 = 10\ c^*$.

Hence, according to equation (6.7.1) the costs are as follows:

$$c = n_1 \cdot c_1 + n_1 \cdot n_2 \cdot c_2 + n_1 \cdot n_2 \cdot n_3 \cdot c_3 + n_1 \cdot n_2 \cdot n_3 \cdot n_4 \cdot c_4$$

$$c = 10 : 3 : 10\ c^* + 10 : 3 : 2 : 10\ c^* + 10 : 3 : 2 : 2 : c^*$$

$$c = 1020\ c^* \qquad (6.8.4)$$

It should be checked whether a better selection could have been made under these assumed conditions:

1) Costs given

Which way of dividing yields the optimum selection for the same costs $c = 1{,}020\ c^*$?

The negligible cost $c_1 \cong 0$ for the selection of the elements of the 1st stage demand that all elements, that is to say all cars, must be included in the investigation, i.e. $n_1 =$

$N_1 = 20$. Hence, by equation (6.8.1):

$$\sigma^2(\bar{x}) = \frac{1}{N_1} \left(\frac{\sigma_2^2}{n_2} + \frac{\sigma_3^2}{n_2 \cdot n_3} + \frac{\sigma_4^2}{n_2 \cdot n_3 n_4} \right) \tag{6.8.5}$$

An optimum selection must now be made for the 3-stage selection 2 - 4. Analogously to equations (6.7.7 a, b):

$$n_4 = \sqrt{\frac{c_3}{c_4}} \cdot \frac{\sigma_4}{\sigma_3}$$

$$n_3 = \sqrt{\frac{c_2}{c_3}} \frac{\sigma_3}{\sigma_2}$$

Hence:

$$n_3 = \frac{0,3 \cdot 10^{-2}}{2,24 \cdot 10^{-2}} < 1$$

$$n_4 = \sqrt{10} \cdot \frac{0,04 \cdot 10^{-2}}{0,3 \cdot 10^{-2}} < 1$$

In deriving the optimum, the n_i were assumed to be continuous. In reality, however, the n_i are discrete and therefore:

$$n_3 = n_4 = 1 \tag{6.8.6}$$

The still unknown number of elements n_2 can be calculated from the total costs (equation 6.7.1):

$$c = 1020\, c^* = N_1 \cdot n_2 \cdot 10\, c^* + N_1 \cdot n_2 \cdot 10\, c^* + N_1 \cdot n_2 \cdot c^*$$

$$n_2 = \frac{1020}{21 \cdot N_1} = 2,43$$

Since the cost c should not be exceeded, let:

$$n_2 = 2 \qquad (6.8.7)$$

Hence, the variance is given by:

$$\sigma^2(\bar{x}) = \frac{1}{20}\left(\frac{5\cdot 10^{-4}}{2} + \frac{0,09\cdot 10^{-4}}{2} + \frac{0,0016}{2}\cdot 10^{-4}\right)$$

$$\sigma^2(\bar{x}) = 0,13 \cdot 10^{-4} \qquad (6.8.8)$$

For a confidence coefficient of 95 %, the mean ash content therefore has a confidence interval of

$$\mu = \bar{x} \pm 1 \cdot 10^{-2} \qquad (6.8.9)$$

with the total cost being

$$c = 20 \cdot 2 \cdot 10\, c^* + 20 \cdot 2 \cdot 10\, c^* + 20 \cdot 2 \cdot c^*$$

$$c = 840\, c^* \qquad (6.8.10)$$

Therefore, a markedly increased precision is achieved even for less expense.

2. Variance given

If minimum possible cost is to be achieved for a given precision, then the selection stages are calculated analogously. If it is assumed th`t, for a confidence coefficient of 95 %, a confidence interval is required which is given by:

$$\mu = \bar{x} \pm 0,5 \cdot 10^{-2} \qquad (6.8.11)$$

Hence:

$$\sigma^2(\bar{x}) = 0,07 \cdot 10^{-4} \qquad (6.8.12)$$

The considerations for n_1, n_3 and n_4 are analogous to the calculations carried out for the case where the cost was given:

$$n_1 = N_1 = 20 \qquad n_3 = n_4 = 1 \qquad (6.8.13)$$

The still unknown selection n_2 is calculated from equation (6.8.5):

$$n_2 = \frac{\sigma_1^2 + \sigma_3^2 + \sigma_4^2}{N_1\, \sigma^2(\bar{x})} = 3,64$$

Hence:

$$n_2 = 4 \qquad (6.8.14)$$

and the costs are:

$$c = 20 \cdot 4 \cdot 10\, c^* + 20 \cdot 4 \cdot 10\, c^* + 20 \cdot 4 \cdot c^*$$

$$c = 1680\, c^* \qquad (6.8.15)$$

6.9 Double sampling for bulk material

In assessing materials, the two parties (supplier and customer) as a rule each analyze at least one sample. A customary procedure of multi-stage sampling and evaluation is as follows: a certain number of n_1 individual samples of the same mass g, the size of which depends essentially on the particle size distribution of the material, are taken randomly from the total amount G. These individual samples are combined to give a collective sample. This collective sample is blended thoroughly. The variance of the elements of the second stage can be thus reduced to such an extent that in all further dividing and comminuting processes only one sample has to be taken in each case (Section 6.7), until the analytical samples are obtained. For the sake of simplicity, it is assumed that each party carries out one analysis each with the analytical values x_I and x_{II}. In general, the measured values x_I and x_{II} will not agree with one another. For this reason, there are occasional disputes about the total content of the bulk material delivered. In order to assess this situation, the question must be answered whether the differences measured are significantly different from one another, for a given statistical certainty. To this end an arror analysis is required. If the variance of the attribute values μ_i of all the units of the 1st step is σ_1^2, and if all the errors which occur in preparing the sample, from the collective sample to the analytical sample, are summarized as σ_{Pr}^2, then the variance for the analysis attribute x is given by:

$$\sigma^2(x) = \frac{\sigma_1^2}{n_1} + \sigma_{Pr}^2 + \sigma_A^2 \qquad (6.9.1)$$

G is large compared with $n_1 \cdot g$. σ_A^2 is the analytical error.

There are now three possible cases:

A) Separate sampling

Each party carries out sampling, from taking the n_1 individual samples, via the sample preparation to the analysis, on his own, but they both use exactly the same procedures and hence have the same sources of error. The analytical values x_I and x_{II} are therefore completely independent of one another. For the difference of the two values we have:

$$\sigma_D^2(x_I - x_{II}) = \sigma^2(x_I) + \sigma^2(x_{II}) = 2\left(\frac{\sigma_1^2}{n_1} + \sigma_{Pr}^2 + \sigma_A^2\right) \qquad (6.9.2)$$

The variance σ_D^2 is a measuring unit for the permissible difference. If it is agreed that the difference must not exceed a certain limit, for example $2\,\sigma_D$, then beyond this limit the difference is called "significant". Complaints within the statistical confidence level are therefore justified. If the difference is smaller than the limit given, then the difference is not significant and an improved estimate for the population is obtained by

$$\bar{x} = \frac{1}{2}(x_I + x_{II}) \qquad (6.9.3)$$

with the variance

$$\sigma^2(\bar{x}) = \frac{1}{2}\left(\frac{\sigma_1^2}{n_1} + \sigma_{Pr}^2 + \sigma_A^2\right) \qquad (6.9.4)$$

B) Joint collective sample, separate sample preparation

The two parties jointly obtain the collective sample and thereafter prepare the samples separately but using the same schemes and instruments. The analytical values x_I and x_{II} are hence no longer independent. Rather, the same errors occur in taking the individual samples. These errors are therefore not involved in the difference an thus:

$$\sigma_D^2(x_I - x_{II}) = 2(\sigma_{Pr}^2 + \sigma_A^2) \qquad (6.9.5)$$

Hence, the variance σ_1^2, whichis after all the measuring unit for the distribution of the attribute in the population, is not involved. Good agreement between x_I and x_{II} does not therefore indicate whether the unknown attribute of the population has been measured well or badly. In this procedure, only the preparation of the collective sample and the analysis are checked.

C) Joint sampling and sample preparation, separate analysis

If the two parties carry out the analysis separately, then the difference of the measured values can also only contain information about the measuring error:

$$\sigma_D^2 (x_I - x_{II}) = 2\,\sigma_A^2 \qquad (6.9.6)$$

Example for double sampling, according to Wilrich[10)]

In sampling raw, fine coal, a collective sample was prepared from $n_1 = 20$ individual samples of 48 kg each in order to determine the ash content, and the variances of the mass concentrations were known from prior investigations:

$\sigma_1^2 = 40 \times 10^{-4}$, $\sigma_{Pr}^2 = 0.20 \times 10^{-4}$ and
$\sigma_A^2 = 0.002 \times 10^{-4}$.

a) Separate sampling:

$$\sigma_D^2 = \frac{2}{20} \cdot 40 \cdot 10^{-4} + 2.0,20 \cdot 10^{-4} + 2.0,002 \cdot 10^{-4} = 4,4 \cdot 10^{-4} \qquad (6.9.7)$$

A significant difference of 2 $_D$ is present if $(x_I - x_{II})$ is greater than 4.2×10^{-2}. Otherwise, a confidence interval for the mean $x = 1/2(x_I + x_{II})$ is given with a statistical confidence of P = 95 % by

$$\mu = \bar{x} \pm 2,1 \cdot 10^{-2} \qquad (6.9.8)$$

b) Joint collective sample, separate sample preparation

$$\sigma_D^2 = 2 \cdot 0,2 \cdot 10^{-4} + 2 \cdot 0,002 \cdot 10^{-4} = 0,4 \cdot 10^{-4} \qquad (6.9.9)$$

Hence, a significant difference of 2 σ_D is given for the analytical results if $|x_I - x_{II}| > 1.26 \times 10^{-2}$.

c) Joint sampling, separate analysis

$$\sigma_D^2 = 2 \cdot 0{,}004 \cdot 10^{-4} = 0{,}008 \cdot 10^{-4} \tag{6.9.10}$$

A significant difference exists between the analytical results, if $|x_I - x_{II}| > 0.126 \times 10^{-2}$.

Compared with separate sampling, the variances for sample preparation and for analyses play an evermore important role in joint processing.

7 Sampling from a Random Mixture

7.1 Sampling with a constant number n of particles from a two-material mixture

7.1.1 Variance of the numerical concentration (Table 7.1.1)

It is often assumed or postulated in sampling that populations, and their strata or stages, are in the state of a random mixture. How- ever, as far as an individual case is concerned, the existence of a random mixture, in the literature also frequently called a stochastic homogeneity, has to be proved. This check can only be effected by sampling. Formerly, the urn scheme was used as a model for a random mixture. An urn was a container in which the elements had been mixed "well". That is to say the probability of finding a particle of one kind should be independent of adjacent particles, regardless of the numerical concentration in the total mixture. The random mixture is then checked by collecting several elements in a sample. If all samples are of the same size n, then we obtain for a random mixture the sampling distributions already derived: a hyper-geometric distribution for sampling without replacement and a binomial distribution for sampling with replacement. At this point it should be checked whether the experimentally determined distributions agree with the theoretically expected distributions. However, in practice frequently only the variances are checked, though it is quite conceivable that no random mixture exists despite agreement of the variances. If such a suspicion does exist, then the procedures for checking non-stochastic homogeneous mixtures must be resorted to.

Table 7.1.1 Sampling with a constant number n of particles in the sample from a mixture of materials (numerical concentration)

1st task: Comparison of the experimental variance $s(x)^2$ of the numerical concentration with the variance of a random mixture

1st step Fixing the sample size n

2nd step Fixing the numerical concentration p_i of the component i possibly by estimating the mean of the numerical concentration x_{ij} from l samples. (For a two-material mixture i = 1 and 2).

$$p_i \approx \bar{x}_i = \frac{1}{l} \sum_{j=1}^{l} x_{ij}$$

3rd step Determination of the sample variance of l samples

(2.3.3) $$s^2 = \frac{1}{l} \sum_{j=1}^{l} (x_{ij} - p_i)^2 \quad \text{or}$$

(2.3.2) $$s^2 = \frac{1}{l-1} \sum_{j=1}^{l} (x_{ij} - \bar{x}_i)^2$$

4th step Calculation of the variance of the numerical concentration x_i

(7.2.6) $$\boxed{\sigma^2(x_i) = \frac{p_i(1-p_i)}{n}}$$

5th step Assignment of a prediction level S(e.g. 95%)

6th step Find the solutions χ_u^2 and χ_o^2 from chi-square distribution $F_f(\chi^2)$ with the degree of freedom f = l - 1

(5.3.8) $$\boxed{\begin{aligned} F(\chi_u^2) &= \frac{1}{2}(1 - S) \\ F(\chi_o^2) &= \frac{1}{2}(1 + S) \end{aligned}}$$

7th step The sample variance agrees with the theoretically expected variance at the preassigned prediction level S, if:

(7.1.6) $$\boxed{\frac{\chi_u^2}{l-1} < \frac{s^2}{\sigma^2} \le \frac{\chi_o^2}{l-1}}$$

2nd task: What must the sample size n be, so that the numerical concentration x_i has a certain probability S of deviating by not more than a preassigned amount f_p from the nominal concentration p_i?

1st step Find the solution c from the normal distribution*):

(5.2.4)
$$S = 2\,H(c) - 1$$

2nd step The condition equation for n follows from the condition required for the confidence interval for a single sample (1 = 1), namely $p_i = x_i \pm c \cdot \sigma/\sqrt{1}$ where $c \cdot \sigma/\sqrt{1}$ f_p and the equation (7.2.6) $\sigma^2 = p_i(1-p_i)/n$

$$n \geq \frac{c^2}{f_p^2} \cdot p_i \cdot (1 - p_i)$$

*)The multinomial distribution is approximated by the normal distribution (rule of thumb $n.p_i > 4$, $n(1-p_i) > 4$, otherwise c has to be found from the corresponding values of the multinomial distribution).

In the case of a random mixture, we have for the theoretically expected variance in sampling without replacement the following equation:

$$\sigma^2(n_x) = \frac{N-n}{N-1}\; n \cdot p(1-p) \qquad (7.1.1)$$

and in the case of sampling without replacement:

$$\sigma^2(n_x) = n \cdot p\,(1-p) \qquad (7.1.2)$$

N is the total number of elements, n the sample size and p the numerical concentration in the total mixture. If the numerical concentration of the sample $x = n_x/n$ is introduced into the equations, then the rules for variances give:

$$\sigma^2(x) = \frac{N-n}{N-1} \; \frac{p(1-p)}{n} \tag{7.1.3}$$

or

$$\sigma^2(x) = \frac{p(1-p)}{n} \tag{7.1.4}$$

These theoretically expected variances for random mixtures must be compared with the sample variances from l replications of n elements each:

$$s^2(x) = \frac{1}{l} \sum_{i=1}^{l} (x_i - p)^2 \tag{7.1.5}$$

A confidence interval for sample variances from normally distributed populations can be derived with the aid of the chi-square distribution. Since the binomial distribution can frequently be closely approximated by the normal distribution even under weak conditions (n.p > 4), the confidence interval determined with the aid of the chi-square distribution is usually likewise a sufficiently good approximation. The sample variance agrees at a given prediction level with the theoretically expected variance, if:

$$\frac{\chi_u^2}{l-1} < \frac{s^2}{\sigma^2} \leq \frac{\chi_o^2}{l-1} \tag{7.1.6}$$

The limits χ^2_e and χ^2_u can be read off from the chi-square distribution for the prediction level required.

Example:

The example chosen is the distribution of even and odd numbers in a random number table. Here we have p = 1 - p = 0.5. Two of the numbers which comprise a four digit sequence of numerals are chosen as sample size, that is to say n = 8. The theoretically expected variance is therefore

$$\sigma^2(x) = \frac{p\cdot(1-p)}{n} = \frac{0,25}{8} = 0,031 \qquad (7.1.7)$$

The sample variance is determined from l = 100 samples.

From equation 1.5 we have

$$s^2 = 0,027 \qquad (7.1.8)$$

for an arithmetic mean of $\bar{x} = 0.495$.

If the lower prediction limit is $P_e = 2.5$ % and the upper is $P_o = 97.5$ %, then we have for l = 100 sampels: $\chi^2_u = 74.2$ and $^2_o = 129.6$. The condition 1.6) is satisfied:

$$0,749 < 0,867 \leq 1,310 \qquad (7.1.9)$$

The sample variance thus agrees at a 95 % prediction level with the theoretical variance of a random mixture.

7.1.2 Variance of the concentration by mass (Table 7.1.2)

Sample concentrations are very frequently determined not by the number of elements, but by weighing. If we continue to assume that the number n of elements in the sample is constant, then the equation (7.1.4) can be converted to concentration by mass. Let the components consist of two even-grained fractions having individual grain masses g_x and g_y and we then have for the sample concentration by mass:

$$X = \frac{n_x \cdot g_x}{n_x \cdot g_x + n_y \cdot g_y} \tag{7.1.7}$$

The condition that the constant number of elements $n = n_x + n_y$ gives:

$$X = \frac{n_x \cdot g_x}{n_x (g_x - g_y) + n \cdot g_y} \tag{7.1.8}$$

To calculate the variance $\sigma^2(X)$ the approximation formula (2.9.5) is used:

$$\sigma^2(X) = \left(\frac{d\,X(E(n_x))}{d\,n_x}\right)^2 \sigma^2(n_x) \tag{7.1.9}$$

$$\sigma^2(X) = \left(\frac{n \cdot g_x \cdot g_y}{(E(n_x) \cdot g_x + E(n_y) \cdot g_y)^2}\right)^2 \cdot \sigma^2(n_x) \tag{7.1.10}$$

To simplify, the expectation of the sample mass is determined:

$$E(g) = E(n_x) \cdot g_x + E(n_y) \cdot g_y = n(p \cdot g_x + (1-p) \cdot g_y) \tag{7.1.11}$$

The second factor describes an average single-grain mass:

$$\bar{g}_{xy} = p \cdot g_x + (1-p) \cdot g_y \tag{7.1.12}$$

Hence we have from equation (7.1.10) together with $\sigma^2(X)$ from equation (7.1.4):

$$\sigma^2(X) = \frac{g_x^2 \cdot g_y^2}{n^2 \cdot \bar{g}_{xy}^4} \cdot n \cdot p\,(1-p) \tag{7.1.13}$$

Table 7.1.2	Sampling with the a constant number n of particles in the sample from a two-material mixture (concentration by mass)
1st task	Comparison of the experimental variance $s^2(X)$ of the concentration by mass X with the variance of a random mixture
1st step	Fixing the sample size n
2nd step	Fixing the concentration by mass P, possibly by estimating the mean of the concentrations by mass X_j from l samples $$P \approx \overline{X} = \frac{1}{l} \sum_{j=1}^{l} X_j$$
3rd step	Determination of the sample variance of l samples
(2.3.3)	$s^2 = \frac{1}{l} \sum_{j=1}^{l} (X_j - P)^2$ or
(2.3.2)	$s^2 = \frac{1}{l-1} \sum_{j=1}^{l} (X_j - \overline{X})^2$
4th step	Calculation of the variance of the concentration by mass
(7.1.18)	$\sigma^2(X) = \frac{P(1-P)}{E(g)} \cdot (P \cdot g_y + (1-P)\, g_x)$
	E(g) is the expectation of the sample mass g for sampling a constant number n of particles. g_x and g_y are the single-grain masses of the components (X) and (Y).
5th step	Assignment of a prediction level S(e.q. 95%)
6th step	Find the solutions χ_u^2 and χ_o^2 from the chi-square distribution $F_f(\chi^2)$ with the degree of freedom f = l - 1
(5.3.8)	$F(\chi_o^2) = \frac{1}{2}(1 - S)$ $F(\chi_u^2) = \frac{1}{2}(1 + S)$

7th step The sample variance agrees with the theoretically expected variance at the preassigned prediction level S, if:

(7.1.6)
$$\frac{\chi_u^2}{l-1} < \frac{s^2}{\sigma^2} \leq \frac{\chi_o^2}{l-1}$$

<u>2nd task:</u> What size must the average sample mass E(g) (for an assumed constant number n of particles), so that the concentration by mass X has a certain probability S of not deviating by more than a preassigned amount f_p from the nomial concentration P?

1st step Find the solution c from the normal distribution*):

(5.2.4)
$$S = 2\,H(c) - 1$$

2nd step The conditional equation for E(g) follows from the condition required for the confidence interval of a single sample (l = 1), $P = X \pm c \cdot \sigma/\sqrt{l}$ where $c \cdot \sigma/\sqrt{l}$ f_p and the equation (7.1.18):

$$E(g) \geq \frac{c^2}{f_p^2} \; P(1-P)\,(P \cdot g_y + (1-P)\,g_x)$$

*)It is assumed that the distribution is distributed approximately normally.

A relationship analogous to the equation (7.1.8) exists between the numerical concentration in the total mixture p and the concentration by mass P:

$$P = \frac{p \cdot g_x}{p \cdot g_x + (1-p) \cdot g_y} \qquad (7.1.14)$$

With the mean sample mass from equation (7.1.12) we have:

$$P = \frac{p \cdot g_x}{\bar{g}_{xy}} \quad \text{und } (1-P) = \frac{(1-p) \cdot g_y}{\bar{g}_{xy}} \qquad (7.1.15)$$

or

$$p = \frac{\bar{g}_{xy}}{g_x} \cdot P \text{ und } (1-p) = \frac{\bar{g}_{xy}}{g_y} \cdot (1-P) \qquad (7.1.16)$$

Hence we have:

$$\frac{\bar{g}_{xy}}{g_x} \cdot P + \frac{\bar{g}_{xy}}{g_y} (1-P) = 1$$

$$\frac{1}{\bar{g}_{xy}} = \frac{P}{g_x} + \frac{(1-P)}{g_y} \qquad (7.17)$$

Using equations (7.1.16) and (7.1.17), the variance of the concentration by mass X for a constant number of elements in the sample is given by:

$$\boxed{\sigma^2(X) = \frac{P \cdot (1-P)}{E(g)} \cdot (P \cdot g_y + (1-P) \cdot g_x)} \qquad (7.1.18)$$

K. Stange[11)] was the first to give the relationship in this form.

7.2 Sampling with a constant sample size n from a multi-material mixture (Table 7.1.1)

Frequently, the mixture to investigated does not consist of two, but of several components. In this case, just as for the two-material mixture, in the model of the urn scheme, the probability of finding a particle of the type j is independent of adjacent particles and equal to the numerical concentration p_j in the total mixture. If an element is removed from the urn and its character noted, and the element is then put back (sampling with replacement!), then the probabilities of finding exactly n_1 elements of the type (1), n_2 elements of the type (2) etc in the n elements form the multinomial distribution which is analogous to the binomial distribution:

$$P_M(n_1, n_2, n_3, \dots n_m) = \frac{n!}{n_1! n_2! \dots n_m!} \cdot p_1^{n_1} \cdot p_2^{n_2} \dots p_m^{n_m} \tag{7.2.1}$$

Additional conditions are:

$$n_1 + n_2 + \dots n_m = n \tag{7.2.2}$$

$$p_1 + p_2 + \dots p_m = 1 \tag{7.2.3}$$

The variance of the individual number n_j is given by[12)]

$$\sigma^2(n_j) = n \cdot p_j(1 - p_j) \tag{7.2.4}$$

Because of the condition (7.2.2) the components are not independent. There is therefore a covariance[12)]:

$$\mathrm{Cov}\,(n_i, n_j) = - n \cdot p_i \cdot p_j \tag{7.2.5}$$

The variance, which is important in practice, of the numerical concentration x_j follows directly from the equation (7.2.4):

$$\boxed{\sigma^2(x_j) = \frac{p_j(1-p_j)}{n}} \tag{7.2.6}$$

In the case of just two components, the multinomial distribution is reduced to the binomial distribution and equation (7.2.6) is identical to equation (7.1.4).

7.3 Sampling with a constant mass (or volume) from a two-component mixture

7.3.1 The problem of a random mixture formed from two even-grained fractions of any size

In practice, sampling with a constant number of particles can be realised only very rarely and sampling with a constant (or almost constant) sample mass g is much more common. For a mixture of two even-grained fractions which do not differ in the masses of the individual grains but in some other property (for example colour) all samples of a constant mass also contain the same number. The variance of the sampling distribution with X_1 as the concentration by mass and g_1 as the mass of a single grain follows directly, with the sample mass g, from the variance of the numerical concentration (equation 7.1.3) or (7.1.4) for sampling without replacement:

$$\sigma^2(X_1) = \frac{G-g}{G-g_1} \cdot P_1(1-P_1) \cdot \frac{g_1}{g} \qquad (7.3.1)$$

and for sampling with replacement:

$$\sigma^2(X_1) = P_1(1-P_1) \cdot \frac{g_1}{g} \qquad (7.3.2)$$

G is the total mass of the mixture and P_1 is the nominal concentration by mass of the component (1).

Solution by Lacey[13], Gy[14], Stange[11], Poole, Taylor and Wall[15], and Kristensen[16] for mixtures of particles of different single-grain masses g_i always also assume that the constant sample mass g contains an (approximately) constant number.

The effects of single-grain weights differing is only accounted for in an average single-grain mass. However, in practice the number cannot be assumed to be constant. Even if the ratio of the diameters of the grain fractions to be mixed is just $d_1/d_2 = 1.2$, and assuming that the components have the same density and that the sample mass corresponds to a mean number of particles of n = 150, a range of between 100 and 200 particles is possible. However, the diameter range of 1.2 which is considered as even-grained in most practical problems encountered in the materials processing industry. The formulae of the authors mentioned are therefore valid only for almost identical single-grain masses.

It will be shown later that the rigorous and exact adherence to the sampling condition of "constant sample mass" is industrially very unrealistic for most grain mixtures. For the purpose of constructing a model, this condition shall nevertheless be maintained. Then we have for the sample masses:

$$g = n_1 \cdot g_1 + n_2 \cdot g_2$$
$$n = n_1 + n_2 \tag{7.3.3}$$

n is the total number of particles, and n_1 and n_2 are the number of particles of the respective components. Since g_1 and g_2 are assumed to be different, not all positive integers between 0 and n are possible for n_1 and n_2 under the condition of an exactly constant sample mass. For two different samples I and II we have for the sample I:

$$g = n_{1\,I} \cdot g_1 + n_{2\,I} \cdot g_2$$
$$n_I = n_{1\,I} + n_{2\,I} \tag{7.3.4}$$

and for the sample II:

$$g = n_{1\,II} \cdot g_1 + n_{2\,II} \cdot g_2$$
$$n_{II} = n_{1\,II} + n_{2\,II} \qquad (7.3.5)$$

Hence:

$$(n_{1\,I} - n_{1\,II})\left(\frac{g_1}{g_2} - 1\right) = n_{II} - n_I \qquad (7.3.6)$$

Let the ratio g_1/g_2 be represented by a cancelled, common fraction:

$$g_1/g_2 = \frac{i}{j} \qquad (7.3.7)$$

i and j are integers without a common factor. Hence the equation (7.3.6) can be rewritten as:

$$(n_{1\,I} - n_{1\,II}) \cdot \left(\frac{i-j}{j}\right) = n_{II} - n_I \qquad (7.3.8)$$

Since $n_{II} - n_I$ is an integer, $(n_{1I} - n_{1II})$ must contain the number j as divisor. If we assume that it should be possible to find only particles of the component (1) or only particles of the component (2) in the sample, then we have for the possible numbers of particles of the components (1) and (2):

$$n_1 = \mu \cdot j \qquad \mu = 0, 1, 2, \ldots$$
$$n_2 = \nu \cdot i \qquad \nu = 0, 1, 2, \ldots \qquad (7.3.9)$$

With equation (7.3.3):

$$g = \mu \cdot j \cdot g_1 + \nu \cdot i \cdot g_2$$
$$\frac{g}{j \cdot g_1} = \mu + \nu \qquad (7.3.10)$$

Since not only μ but also λ are positive integers, $g/j \cdot g_1 = \lambda$ must also be a positive integer. The condition of a constant sample mass therefore only makes the sense for discrete values of g.

Hence we have:

$$\lambda = \mu + \nu \qquad \text{mit } \lambda = \text{const} \qquad (7.3.11)$$

The derivations make it quite clear that an unmodified urn-scheme is no longer a suitable model for the problem of sampling with a constant mass from a random mixture, because the probability of an element being included in a sample is no longer independent of the elements which are already in the sample. Rather, the selection rules (7.3.10) have to be taken into account. That is to say it is only possible for units of the size $j \cdot g_1$ or $i \cdot g_2$ to be exchanged with one another. The urn scheme therefore has to be modified: a random mixture is said to exist in this case if the probability that a unit contains j particles of the type (1) is independent of other units regardles of the concentration P_1 in the total mass.

It does not matter in this case what the spatial relationship of j particles is which form a unit.

Under these conditions we have for the numbers μ, for constant , a hypergeometric or a binomial distribution, analogous to earlier sampling distributions:

$$P_B(\mu) = \binom{\nu}{\mu} P_1^{\mu} (1 - P_1)^{\nu} \qquad (7.3.12)$$

The variance $\sigma^2(\mu)$ of the binomial distribution is

$$\sigma^2(\mu) = P_1(1 - P_1) \cdot \lambda \qquad (7.3.13)$$

It follows that the sample mass concentration X_1 is given by:

$$X_1 = \frac{n_1 \cdot g_1}{g} = \frac{n_1/j}{g/j \cdot g_1} = \frac{\mu}{\lambda} \qquad (7.3.14)$$

Using the rules for variances the sample mass concentration is then given by:

$$\boxed{\sigma^2(X_1) = P_1(1-P_1) \cdot \frac{g_1 \cdot j}{g}} \qquad (7.3.15)$$

This is a mathematically exact result, but it cannot be used in this form in practice. A simple example will show this: if in one case there is a mass ratio of two single grains of $g_1/g_2 = 11/10$, and in the other case of $g_1/g_2 = 111/100$, then the corresponding variances, with $j_1 = 10$ and $j_2 = 100$, are in theory, for the same sample weight, related to one another as 1 : 10. However, this result is not expected for the mixtures, which in practice are almost identical. The discrepancy between theory and practice arises in the postulated condition of an exactly constant sample mass and the selection rule, usually violated experimentally, of $g = \lambda \cdot j \cdot g_1$.
Since the actually present variations in sampling are not observed, no statements about the true ratios are possible in the general case. However, the important special case in which a grain mass of one component is an integral multiple of the grain mass of the other component does yield important results which can be used in practice.

7.3.2 Random mixture of two even-grained fractions, where the grain mass of one component is an integral multiple of the grain mass of the other component (Table 7.3.1)

The assumption made here that the grain mass of the component (1) is an integral multiple of the grain mass (2) of the other

Table 7.3.1 Sampling with a constant sample mass g from a random mixture of two even-grained fractions having single-grain masses g_1 and g_2

1st task	Comparison of the experimental variance $s^2(X_1)$ of the concentration by mass X_1 with the variance of the random mixture
1st step	Fixing the sample size g
2nd step	Fixing the concentration by mass P_1, possibly by estimating the mean of the concentration by mass X_{1j} from l samples $P_1 \approx \bar{X}_1 = \frac{1}{l} \sum_{j=1}^{l} X_{1j}$
3rd step	Determination of the sample variance of l samples
(2.3.3)	$s^2 = \frac{1}{l} \sum_{j=1}^{l} (X_{1j}-P)^2$ or
(2.3.2)	$s^2 = \frac{1}{l-1} \sum_{j=1}^{l} (X_{1j}-\bar{X}_1)^2$
4th step	The assumption is made that the grain mass g_1 is an integral multiple of the grain mass g_2. The smallest possible derivation of the fraction is thus $d_1/d_2=\sqrt[3]{2}$ 1.26
5th step	Calculation of the variance of the concentration by mass X_1
(7.3.17)	$\sigma^2(X_1) = P_1(1-P_1) \cdot \frac{g_1}{g}$
6th step	Assignment of a prediction level S(e.q. 95%)
7th step	Find the solutions χ_u^2 and χ_o^2 from chi-square distribution $F_f(\chi^2)$ with the degree of freedom $f = l - 1$
(5.3.8)	$F(\chi_o^2) = \frac{1}{2}(1-S)$ $F(\chi_u^2) = \frac{1}{2}(1+S)$

8th step The sample variance agrees with the theoretically expected variance at the preassigned prediction level S, if:

$$\frac{\chi_u^2}{l-1} < \frac{s^2}{\sigma^2} \leq \frac{\chi_o^2}{l-1}$$

<u>2nd task:</u> What must the size of the sample mass be so that the concentration by mass X_1 has a certain probability S of not deviating by more than a preassigned amount f_p from the nomial concentration P_1?

1st step Find the solution c from the normal distribution*):

$$S = 2\,H(c) - 1 \tag{5.2.4}$$

2nd step The conditional equation for g follows from the condition required for the confidence interval for a single sample (l = 1), namely $P_i = X_i \pm c \cdot \sigma/\sqrt{l}$ where $c \cdot \sigma/\sqrt{l} \leq f_p$ and the equation (7.3.17):

$$g \geq \frac{c^2}{f_p^2} \cdot P_1(1-P_1) \cdot g_1$$

*) It is assumed that the distribution is distributed approximately normally.

component is not only a theoretically important special case but also of practical importance. The smallest possible ratio of this kind is just $g_1/g_2 = 2$. Thus the smallest particle-size ratio is, for identical density, $d_1/d_2 = \sqrt[3]{2} = 1.26$. However, in practice a diameter ratio of $d_1/d_2 = 1.2$ is still referred to as even-grained. With the assumptions made, the equations simplify. From equation (7.3.9) we have, for j = 1:

$$n_1 = \mu \qquad \mu = 0, 1, 2 \ldots$$
$$n_2 = \nu \cdot i \qquad \nu = 0, 1, 2 \ldots \tag{7.3.16}$$

The number n_1 of the component (1) can thus assume all values between 0 and n. The number n_2 can only change in units of size i. It follows from the equation (7.3.15) that the variance of the weight concentration is given by:

$$\boxed{\sigma^2(X_1) = P_1(1-P_1) \cdot \frac{g_1}{g}} \qquad (7.3.17)$$

The variance contains as the variable only the largest single grain. From the equation (7.3.10) the selection condition is obtained:

$$g = \lambda \cdot g_1 \qquad (7.3.18)$$

In this special case likewise, not all sample sizes are possible but only those which are integral multiples of g_1. The effect of a deviation may be investigated in an example. Here the grain mass g_1 is exactly three times as large as the grain mass of g_2. The result is shown in Figure 7.3.1. For all sample masses which are an integral multiple of the weight g_1, the expectation $E(X_1)$ is identical to the mixture composition P_1 and the variance $\sigma^2(X_1)$ is identical to the variance calculated in accordance with equation (7.3.17). All the other possible sample masses do not give unbiased estimates for P_1 and they give variances which deviate from the variances in accordance with equation (7.3.17). However, the deviations rapidly become negligibly small with increasing sample masses. For large samples the formulae derived are also approximately true for sample masses which are not an exact multiple of the largest single-grain mass.

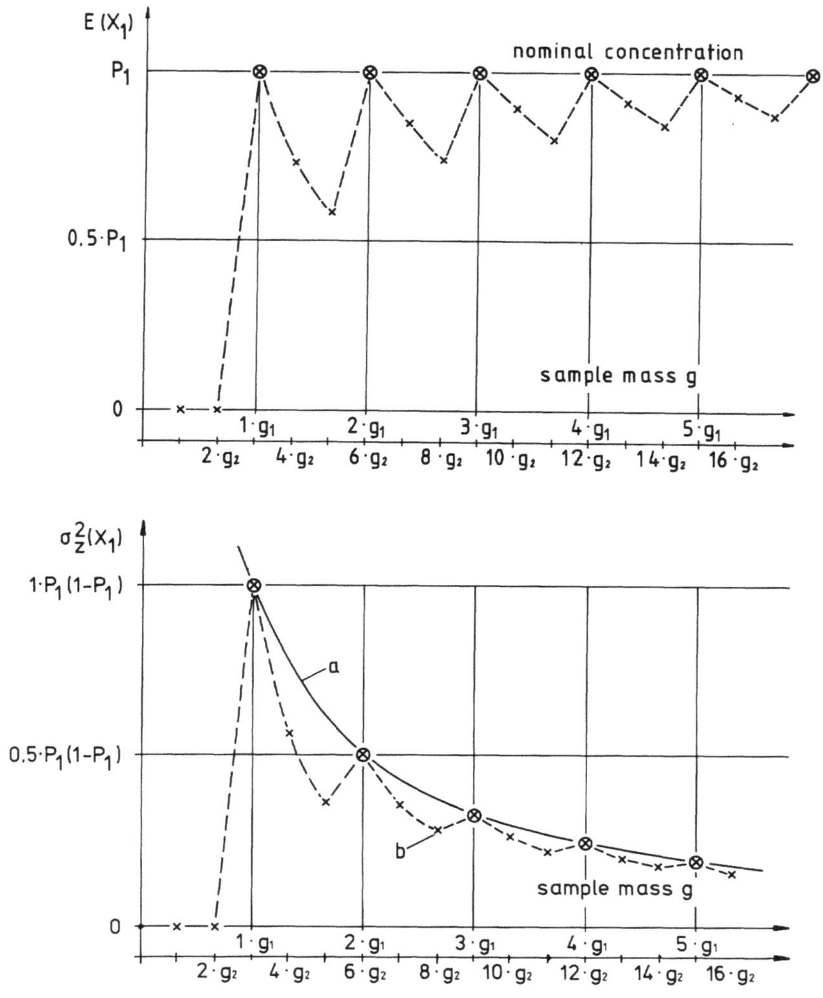

<u>Figure 7.3.1</u> The dependence of the expectation $E(X_1)$ and the variance $\sigma_z^2(X_1)$ of a random mixture on the sample mass g in the special case of a binary mixture having $g_1 = 3g_2$. a = function in accordance with equation (7.3.17) b ≙ calculated values for sample masses which are not an integral multiple of the largest single grain mass g_1.

7.4 Constant sample mass and the random mixture of several components of unequal grain size (Table 7.4.1)

Frequently the mixture to be investigated does not consist of just one but of several components. It is assumed that each component is present in a narrow grain fraction having the single-grain mass g_1. It is also assumed that the single-grain masses g_i are an integral multiple of the respective next smaller single-grain mass g_{i+1}:

$$g_1 = \mu_2 \cdot g_2 = \mu_2 \cdot \mu_3 \cdot g_3 = \dots \qquad \mu_i = 2, 3, \dots \qquad (7.4.1)$$

g_1 is here the heaviest single grain and g_n is the lightest.

To investigate whether such a mixture is a true random mixture, samples of constant sample mass g are also taken in this case. If the number of single grains of the i-th fraction is equal to n_i, then the sample mass is given by:

$$g = n_1 \cdot g_1 + n_2 \cdot g_2 + \dots n_k \cdot g_k + \dots n_n \cdot g_n \qquad (7.4.2)$$

For calculating the variance of the 1st component having the largest single-grain mass, all the other components are put together. The variance of the concentration by mass X_1 is then given for the two-material mixture by Formula (7.3.17):

$$\sigma^2(X_1) = P_1(1 - P_1) \cdot \frac{g_1}{g} \qquad (7.4.3)$$

Here, as below, P_i is the concentration by mass of the i-th component in the total and X_i is the concentration by mass of the i-th component in the sample. It is noteworthy that here the variance is not a function of an average single-grain mass as is the case for the abovementioned authors. The variance only depends on the largest single-grain mass g_1. Equation (7.4.3) is therefore not valid for the k-th component, since g_k is not the largest single-grain mass. Initially it is only possible to make statements about the mixture of the k-th fraction with the

Table 7.4.1	Sampling with a constant sample mass g from a random mixture of several components having single-grain masses g_i.
	Comparison of the experimental variances $s^2(X_i)$ or $s^2(\Sigma X_i)$ of the concentration by mass X_i or ΣX_i with the variances of the random mixture
1st step	Fixing the sample size g and the single grain size g_i with the assumption that g_i is an integral multiple of g_{i+1}. The smallest possible gradation of the fractions thus $\frac{d_i}{d_i+1} = \sqrt[3]{2} \approx 1{,}26.$
2nd step	Fixing the concentration by mass P_i, possibly by estimating the mean of the concentration by mass X_{ij} from l samples $P_i = X_i = \Sigma X_{ij}/l$
3rd step	Determination of the sample variances of l samples
(2.3.3)	$s^2 = \frac{1}{l}\sum_{j=1}^{l}(X_{ij}-P_i)^2$ bzw. $s^2 = \frac{1}{l}\sum_{j=1}^{l}\left(\sum_{i=1}^{k}X_{ij}-\sum_{i=1}^{k}P_i\right)^2$
	or
(2.3.2)	$s^2 = \frac{1}{l-1}\sum_{j=1}^{l}(X_{ij}-\overline{X}_i)^2$ bzw. $s^2 = \frac{1}{l-1}\sum_{j=1}^{l}\left(\sum_{i=1}^{k}X_{ij}-\sum_{i=1}^{k}\overline{X}_i\right)^2$
4th step	Calculation of the variance of the concentration by mass
(7.4.10)	$\sigma^2(X_k) = P_k'(1-P_k')\cdot\frac{g_k}{g}\left(1-\sum_{i=1}^{k-1}P_i\right) + P_k'^2\cdot\sigma^2\left(\sum_{i=1}^{k-1}X_i\right)$
(7.4.16)	$\sigma^2\left(\sum_{i=1}^{k}X_i\right) = P_k(1-P_k')\cdot\frac{g_k}{g} + (1-P_k')^2\cdot\sigma^2\left(\sum_{i=1}^{k-1}X_i\right)$
(7.4.5)	$P_k' = \frac{P_k}{1-\sum_{i=1}^{k}P_i}$
	The variances of the individual components and the variance of the sum of the first k components can thus be calculated recursively for a true random mixture.

5th step		Assignment of a prediction level S(e.q. 95%)
6th step		Find the solutions χ_u^2 and χ_o^2 from chi-square distribution $F_f(\chi^2)$ with the degree of freedom $f = l - 1$
	(5.3.8)	$$F(\chi_u^2) = \frac{1}{2}(1-S)$$ $$F(\chi_o^2) = \frac{1}{2}(1+S)$$
7th step		The sample variance agrees with the theoretically expected variance at the preassigned prediction level S, if :
		$$\frac{\chi_u^2}{l-1} < \frac{s^2}{\sigma^2} \leq \frac{\chi_o^2}{l-1}$$

fractions of smaller single-grain masses:

$$g' = g - \sum_{i=1}^{k-1} n_i g_i = n_k g_k + n_{k+1} g_{k+1} + \dots n_n g_n \tag{7.4.4}$$

If it is assumed for the moment that $\sum_{i=1}^{k-1} n_i g_i$ is constant then g is also constant. This remainder mass g is now divided between the k-th component and the components which have smaller single-grain masses on average in the ratio of their concentrations in the total. It follows that the concentration X_k' of the k-th component and the expectation P_k', each relative to the remainder volume g', are thus given by:

$$X_k' = \frac{X_k}{1 - \sum_{i=1}^{k-1} X_i} \quad ; \quad P_k' = \frac{P_k}{1 - \sum_{i=1}^{k-1} P_i} . \tag{7.4.5}$$

The conditioNal expectation $E(X_k \sum_{i=1}^{k-1} X_i)$ is given by:

$$E(X_k \mid \sum_{i=1}^{k-1} X_i) = (1 - \sum_{i=1}^{k-1} X_i) P_k' \tag{7.4.6}$$

E(X Y) identifies the expectation of the value under the "condition" that Y assumes any fixed value.

Under the condition that $\sum_{i=1}^{k-1} X_i$ and hence also g are constant, the remainder mixture can again be converted into a two-component mixture and, analogously to equation (7.4.3), the conditional variance is given by:

$$\sigma^2(X_k' \mid \sum_{i=1}^{k-1} X_i) = P_k'(1 - P_k') \cdot \frac{g_k}{g'} \qquad (7.4.7)$$

Together with equations (7.4.4) and (7.4.5) it follows that:

$$\sigma^2(X_k \mid \sum_{i=1}^{k-1} X_i) = P_k'(1 - P_k') \cdot \frac{g_k}{g} \cdot (1 - \sum_{i=1}^{k-1} X_i) \qquad (7.4.8)$$

Should the sum $\sum_{i=1}^{k-1} X_i$ likewise be variable, then it follows from the rules for working out conditional expectations and variances that:

$$\sigma^2(X_k) = E\left\{\sigma^2(X_k \mid \sum_{i=1}^{k-1} X_i)\right\} + \sigma^2\left\{E(X_k \mid \sum_{i=1}^{k-1} X_i)\right\} \qquad (7.4.9)$$

The required variance $\sigma^2(X_k)$ is then obtained with the equations (7.4.6) and (7.4.8):

$$\boxed{\sigma^2(X_k) = P_k'(1 - P_k') \cdot \frac{g_k}{g} (1 - \sum_{i=1}^{k-1} P_i) + P_k'^2 \cdot \sigma^2 (\sum_{i=1}^{k-1} X_i)} \qquad (7.4.10)$$

The variance $\sigma^2(\sum_{i=1}^{k-1} X_i)$ is initially still unknown. The following expansion is therefore carried out:

$$\sigma^2(\sum_{i=1}^{k} X_i) = \sigma^2(X_k + \sum_{i=1}^{k-1} X_i) \qquad (7.4.11)$$

$$\sigma^2(\sum_{i=1}^{k}(X_i)) = \sigma^2(X_k) + \sigma^2(\sum_{i=1}^{k-1} X_i) + 2\,E(X_k \cdot \sum_{i=1}^{k-1} X_i) - 2\,P_k \cdot \sum_{i=1}^{k-1} P_i$$

Using the rule for calculating conditional expectations

$$E(X) = E\,\{E(X|\,Y)\} \qquad (7.4.12)$$

it follows that:

$$E(X_k \cdot \sum_{i=1}^{k-1} X_i) = E\left\{E(X_k \cdot \sum_{i=1}^{k-1} X_i \Big|\, \sum_{i=1}^{k-1} X_i)\right\}$$

$$E(X_k \cdot \sum_{i=1}^{k-1} X_i) = E\left\{\sum_{i=1}^{k-1} X_i \cdot E(X_k \Big|\, \sum_{i=1}^{k-1} X_i)\right\} \qquad (7.4.13)$$

$$\begin{aligned}
E(X_k \cdot \sum_{i=1}^{k-1} X_i) &= E\left\{\sum_{i=1}^{k-1} X_i \cdot (1 - \sum_{i=1}^{k-1} X_i) \cdot P_k'\right\} \\
&= P_k' \cdot E\left\{\sum_{i=1}^{k-1} X_i - (\sum_{i=1}^{k-1} X_i)^2\right\} \\
&= P_k' \cdot \left\{E(\sum_{i=1}^{k-1} X_i) - E(\sum_{i=1}^{k-1} X_i)^2\right\} \\
&= P_k' \cdot \sum_{i=1}^{k-1} P_i - P_k'\left\{\sigma^2(\sum_{i=1}^{k-1} X_i) + (\sum_{i=1}^{k-1} P_i)^2\right\} \\
&= P_k' \cdot \sum_{i=1}^{k-1} P_i \cdot \left\{1 - \sum_{i=1}^{k-1} P_i\right\} - P_k' \cdot \left\{\sigma^2(\sum_{i=1}^{k-1} X_i)\right\}
\end{aligned}$$

$$E(X_k \cdot \sum_{i=1}^{k-1} X_i) = P_k \cdot \sum_{i=1}^{k-1} P_i - P_k' \cdot \sigma^2(\sum_{i=1}^{k-1} X_i) \qquad (7.4.14)$$

The insertion of equation (7.4.14) into equation (7.4.11) gives:

$$\sigma^2\left(\sum_{i=1}^{k} X_i\right) = \sigma^2(X_k) + \sigma^2\left(\sum_{i=1}^{k-1} X_i\right) - 2\,P_k' \cdot \sigma^2\left(\sum_{i=1}^{k-1} X_i\right) \qquad (7.4.15)$$

and, together with equation (7.4.10), a recurrence relation is obtained therefrom:

$$\boxed{\sigma^2\left(\sum_{i=1}^{k} X_i\right) = P_k(1 - P_k') \cdot \frac{g_k}{g} + (1 - P_k')^2 \cdot \sigma^2\left(\sum_{i=1}^{k-1} X_i\right)} \qquad (7.4.16)$$

For a true random mixture the variances of the individual components but also the variance of the sum of the first k components can be calculated recursively by using equation (7.4.16) and equation (7.4.10).

Experiments on a multi-component mixture[17)]

A mixture was prepared from five narrow fractions of glass spheres. 25 samples having a constant sample mass of 29 mg were taken from the mixture. The sample mass concentration X_i could be determined, owing to the spherical shape of the particles, by counting in a particle counter. Table (7.4.1) lists in addition to the sample variances $s^2(\Sigma X_i)$ also the theoretically expected $\sigma^2(\Sigma X_i)$. Granted that the relationship between the sample variance and the theoretical variance can be checked approximately by the chi-square distribution, a prediction level of 95 % yields a prediction interval of:

$$0{,}55\ \sigma^2 \leqq s^2 \leqq 1{,}73\ \sigma^2$$

This condition is satisfied for all residues. The mixture is thus stochastically homogeneous within the prediction level assumed.

Table 7.4.2 Comparison of the observed with the calculated variances in a multi-material mixture

k	g_k	P_k	P'_k	$\sum_{i=1}^{k} P_i$	$\sigma^2(X_k)$	$\sigma^2(\sum_{i=1}^{k} X_i)$	$s^2(\sum_{i=1}^{k} X_i)$
1	0,52 mg	0,22	0,22	0,22	$31 \cdot 10^{-4}$	$31 \cdot 10^{-4}$	$34 \cdot 10^{-4}$
2	0,20 mg	0,21	0,27	0,43	$13 \cdot 10^{-4}$	$27 \cdot 10^{-4}$	$21 \cdot 10^{-4}$
3	0,089 mg	0,23	0,40	0,66	$8,7 \cdot 10^{-4}$	$13 \cdot 10^{-4}$	$7,8 \cdot 10^{-4}$
4	0,012 mg	0,12	0,35	0,78	$2,1 \cdot 10^{-4}$	$6,3 \cdot 10^{-4}$	$4,3 \cdot 10^{-4}$
5	0,0018 mg	0,22	1,00	1,00	$5,9 \cdot 10^{-4}$	0	0

k = grain class

g_k = single-grain mass

P_k = mixture composition

P'_k = concentration in the remainder mixture (equation 7.4.5)

$\sigma^2(X_k)$ = variance of the k-grain class (equation 7.4.10)

$\sigma^2(\sum_{i=1}^{k} X_i)$ = variance of the sum of the first k components (equation 7.4.16)

$s^2(\sum_{i=1}^{k} X_i)$ = sample variance from 25 individual samples

7.5 Random mixture of two particle-size distributions (Table 7.5.1)

In the case that we have considered up to now, the components of the mixture were even-grained fractions of differing size. In practice it is much more frequent however that each component (for example overburden and ore) has a grain-size distribution of its own. Two limiting cases will be considered. On the one hand, the grain-size distributions of the components to be mixed are to be the same and, on the other hand, the smallest grain of the component (X) shall be larger than the largest grain of the component (Y). If in the latter case the number of single grains of the i-th fraction is $n_{i,x}$ for the component (X) and $n_{j,y}$ for the component (Y), then the sample mass g is made up of:

$$g = n_{1,x} \cdot g_1 + n_{2,x} \cdot g_2 + \ldots n_{k,x} \cdot g_k + n_{k+1,y} \cdot g_{k+1} + n_{k+2,y} \cdot g_{k+2} \ldots n_{n,y} \cdot g_n \tag{7.5.1}$$

Table 7.5.1	Sampling with a constant sample mass g from a random mixture of two grain-size distributions
Case A:	The smallest grain of the component (X) is larger than the largest grain of the component (Y)
1st task:	Comparison of the experimental variance $s^2(X)$ of the concentration by mass X with the variance of the random mixture
1st step	Fixing the sample mass g
2nd step	Fixing the concentration by mass P of the component (X) and its grain-size screenings distribution Q_3^x, possibly by estimating with the mean from several experiments. $g(Q_3^x)$ is the dependence of the single-grain mass of the component (X) on the screening Q_3^x
3rd step	Determination of the sample variances of l samples $s^2(X) = \frac{1}{l} \sum_{j=1}^{l} (X_j - P)^2$ or $s^2(X) = \frac{1}{l-1} \sum_{j=1}^{l} (X_j - \bar{X})^2$ and $\bar{X} = \frac{1}{l} \sum_{j=1}^{l} X_j$
4th step	Calculation of the variance of the concentration by mass $\boxed{\sigma^2(X) = \frac{Q^2 P}{g} \cdot \int_0^1 \frac{g(Q_3^x)}{(Q - P\,Q_3^x)^2}\, dQ_3^x}$ $\quad Q = 1 - P$

5th step — Assignment of a prediction level S(e.q. 95%)

6th step — Find the solutions χ_u^2 and χ_o^2 from chi-square distribution $F_f(\chi^2)$ with the degree of freedom $f = l - 1$

(5.3.8)
$$F(\chi_u^2) = \frac{1}{2}(1-S)$$
$$F(\chi_o^2) = \frac{1}{2}(1+S)$$

7th step — The sample variance agrees with the theoretically expected variance at the preassigned prediction level S, if:

$$\frac{\chi^2}{l-1} < \frac{s^2}{\sigma^2} \leq \frac{\chi_o^2}{l-1}$$

<u>2nd task:</u> — What size must the sample mass be, so that the concentration by mass has a certain probability S of not deviating by more than a preassigned amount f_p from the nominal concentration P?

1st step — Find the solution c from the normal distribution*):

(5.2.4)
$$S = 2\,H(c) - 1$$

2nd step — The sample size g is given by the condition required for the confidence interval of a single sample ($l = 1$), namely $P = X \pm c\,\sigma/\sqrt{l}$, where $c \cdot \sigma/\sqrt{l} \leq f_p$ and the equation (7.5.9):

$$g \geq \frac{c^2}{f_p^2} \cdot Q^2 \cdot P \cdot \int_0^1 \frac{g(Q_3^x)}{(Q + P\,Q_3^x)^2}\, d\,Q_3^x \qquad Q = 1 - P$$

*)It is assumed that the distribution is distributed approximately normally.

Case B: The mixture components (X) and (Y) have the same grain-size distribution $Q_3^X = Q_3^Y$

1st task: Comparison of the experimental variance $s^2(X)$ for the concentration by mass X with the variance of the random mixture

1st step — Fixing the sample mass g

2nd step — Fixing the concentration by mass P of the component (X) and the grain-size distribution $Q_3^X = Q_3^Y = Q_3$, or the distribution of the residue $R = 1 - Q_3$. g(R) is the dependence of the single-grain mass of the components (X) and (Y) on the residue R

3rd step — Determination of the sample variances of l samples

$$s^2(X) = \frac{1}{l}\sum_{j=1}^{l}(X_j - P)^2 \text{ or}$$

$$s^2(X) = \frac{1}{l-1}\sum_{j=1}^{l}(X_j - \bar{X})^2 \text{ and } \bar{X} = \frac{1}{l}\sum_{j=1}^{l} X_j$$

4th step — The calculation of the variance of the concentration by mass

$$\sigma^2(X) = \frac{P}{g}\cdot\int_0^1 g(R)\,dR - \frac{2P^2}{g}\int_0^1 (1-R)\int_0^R \frac{g(\xi)}{(1-\xi)^2}\,d\xi\,dR$$

For the special case of a power distribution for Q_3 see equation (7.5.31)

5th step — Assignment of a prediction level S(e.q. 95%)

6th step — Find the solutions χ_u^2 and χ_o^2 from chi-square distribution $F_f(\chi^2)$ with the degree of freedom $f = l - 1$

$$F(\chi_u^2) = \frac{1}{2}(1-S)$$
$$F(\chi_o^2) = \frac{1}{2}(1+S) \qquad (5.3.8)$$

7th step — The sample variance agrees with the theoretically expected variance at the preassigned prediction level S, if:

$$\frac{\chi_u^2}{l-1} < \frac{s^2}{\sigma^2} \leq \frac{\chi_o^2}{l-1}$$

<u>2nd task:</u> What size must the sample mass be, so that the concentration by mass has a certain probability S of not deviating by more than a preassigned amount f_p from the nominal concentration P?

1st step Find the solution c from the normal distribution*):

$$S = 2\,H(c) - 1 \qquad (5.2.4)$$

2nd step The sample size g is given by the condition required for the confidence interval of a single sample (l = 1), namely $P = X \pm c\ \sigma/\sqrt{l}$, where $c \cdot \sigma/\sqrt{l} \leq f_p$ and the equation (7.5.9):

$$g \geq \frac{c^2}{f_p^2}\left\{P \cdot \int_0^1 g(R)\,dR - 2\,P^2 \int_0^1 (1-R) \int_0^R \frac{g(\xi)}{(1-\xi)^2}\,d\xi\,dR\right\}$$

*)It is assumed that the distribution is distributed approximately normally

The sample mass thus corresponds to equation (7.4.2). The only thing that changes is the notation for the fraction with single-grain masses smaller than g_k. This however means that the problem is principally the same as for the random mixture of several components having differing grain sizes (Chapter 7.6). The concentration by mass X_i and Y_i are again labelled as follows:

$$X_i = \frac{n_{i,x} \cdot g_i}{g} \text{ und } Y_j = \frac{n_{k+j,y} \cdot g_{k+j}}{g} \qquad (7.5.2)$$

and

$$P_i = E(X_i) \text{bzw. } Q_j = E(Y_j) \qquad (7.5.3)$$

P_i and Q_j are the proportions of the fractions in the total mixture. If P is the proportion by mass of the component (X) in the total mixture and Q = 1 - P is the proportion by mass of the component (Y) then:

$$E(\sum_{i=1}^{k} X_i) = E(X) = P$$

The concentration by mass X is an unbiased estimate for the concentration in the total. We are interested in the variance $\sigma^2(X) = \sigma^2(\Sigma X_i)$ of this sample concentration. By means of the equation (7.4.16) or (7.4.10) a recursive calculation yields:

$$\sigma^2(X) = \sigma^2(\sum_{i=1}^{k} X_i) = P_k(1-P_k') \cdot \frac{g_k}{g} + (1-P_k')^2 \cdot \sigma^2(\sum_{i=1}^{k-1} X_i) \qquad (7.5.4)$$

If the discrete fractions are replaced by continuous distributions, then the following relationships hold:

$$\text{a)} \quad \sum_{i=1}^{j} X_i = R^*(g_j)$$

$$\text{b)} \quad \sum_{i=1}^{j} P_i = R(g_j)$$

$$\text{c)} \quad X_j = dR^* \qquad (7.5.5)$$

$$\text{d)} \quad P_j = dR$$

If it is assumed that dR is relatively small compared with 1 then equation (7.4.16) can be converted into a differential equation:

$$\sigma^2(R^*(g_j)) - \sigma^2(R^*(g_j) - dR^*) = dR \cdot \frac{g_j}{g} - \frac{2\,dR}{1-R} \cdot \sigma^2(R^* - dR^*) \qquad (7.5.6)$$

The solution of the equation yields:

$$\sigma^2(R^*(g_j)) = \frac{(1-R)^2}{g} \int_0^R \frac{g(\xi)}{(1-\xi)^2}\, d\xi$$

Für j → k geht $R^*(g_j) \rightarrow X$ und $R(g_j) \rightarrow P$. Damit gilt:

$$\sigma^2(X) = \frac{(1-P)^2}{g} \int_0^P \frac{g(R)}{(1-R)^2}\, dR \tag{7.5.7}$$

g(R) is the distribution of the grain masses in the total mixture. If $Q_3(\delta)$ is the cumulative grain-size distribution of the component (X), where δ is the dispersity value, then it follows for the case of constant density:

$$Q_3^x(\delta) = (1 - \frac{R(\delta)}{P}) \tag{7.5.8}$$

Using this relationship, the integral in equation (7.5.7) can be transformed to give:

$$\boxed{\sigma^2(X) = \frac{Q^2 P}{g} \int_0^1 \frac{g(Q_3^x)}{(Q + P\, Q_3^x)^2}\, dQ_3^x} \tag{7.5.9}$$

If, as a special case, the component (X) is an even-grain fraction of constant single-grain mass g_1, then it follows from equation (7.5.9):

$$\sigma^2(X) = \frac{g_1}{g} Q^2 \cdot P \int_0^1 \frac{1}{(Q+P\, Q_3^x)^2}\, dQ_3^x = \frac{g_1}{g} Q^2 \left[- \frac{1}{(Q+P\, Q_3^x)} \right]_0^1$$

$$\sigma^2(X) = \frac{g_1}{g} Q\,(1 - Q) \tag{7.5.10}$$

This result agrees, as was to be expected, with the variance of the 1st component of a multi-material mixture (equation (7.4.3).

It is worth noting that only the particle size distribution Q_3^x of the "larger" component enters into the variance $\sigma^2(X)$. If the distributions overlap, that is to say, if the components (X) and (Y) have particles in the same grain-size region, then the grain-size distribution of the component (Y) in the region of overlap can also have an effect. In a special case, the components to be mixed have the same grain-size distribution $Q_3^{X}(\delta)$ and $Q_3^{Y}(\delta)$. Again δ is the dispersity value. If the two components (X) and (Y) are mixed in the weight proportions P and Q = 1 - P, then not only the component (X) but also the component (Y) has particles of identical grain mass g_i. If the number of individual grains of the i-th fraction is $n_{i,x}$ for the component (X) and $n_{i,y}$ for the component (Y) then the sample mass g is made up of:

$$g = n_{1,x} \cdot g_1 + n_{1,y} \cdot g_1 + n_{2,x} \cdot g_2 + n_{2,y} \cdot g_2 + \ldots n_{n,x} \cdot g_n + n_{n,y}\, g_n \tag{7.5.11}$$

Here the same conditions as in section (7.4) are applicable: the single-grain masses g_i are each an integral multiple of the next smaller one. g_1 is the largest individual grain and g_n is the smallest. The following still hold:

$$X_1 + Y_1 + X_2 + Y_2 + \ldots \quad X_n + Y_n = 1 \tag{7.5.12}$$

$$\sum_{i=1}^{n} X_i = X \quad \text{and} \quad \sum_{i=1}^{n} Y_i = Y \tag{7.5.13}$$

$$\sum_{i=1}^{n} P_i = P \quad \text{and} \quad \sum_{i=1}^{n} Q_i = Q \tag{7.5.14}$$

X_i and Y_i are the proportions by mass of the i-th fraction of the respective component of the sample. P_i and Q_i are the corresponding proportions by mass of the i-th fraction of the respective component in the total mixture. X and Y identify the proportion by mass of the components (X) and (Y) in the sample, and P and Q act correspondingly for the total mixture.

The calculation of the variances is analogous to section (7.4). First the variances of the individual fractions are determined by initially only making statements about the mixture of the k-th fraction of a component with the remainder of the same or of smaller grain sizes:

$$g'' = g - \sum_{i=1}^{k-1} n_{i,x} \cdot g_i - \sum_{i=1}^{k-1} n_{i,y} \cdot g_i = n_{k,x} \cdot g_k + n_{k,y} \cdot g_k + n_{k+1,x} g_{k+1} + $$
$$+ n_{k+1,y} \cdot g_{k+1} \; \ldots + n_{n,x} \cdot g_n + n_{n,y} \cdot g_n \qquad (7.5.15)$$

If it is assumed for the moment that $(\sum_{i=1}^{k-1} n_{i,x}\, g_i + \sum_{i=1}^{k-1} n_{i,y} \cdot g_i)$ is constant, then g" is also constant. g" is divided between the k-th fraction of the component (X) and the remainder mixture on average in proportion to their concentrations in the total. This yields for the concentration X''_k of the k-th component:

$$X''_k = \frac{X_k}{1 - \sum_{i=1}^{k-1} X_i - \sum_{i=1}^{k-1} Y_i} \qquad P''_k = \frac{P_k}{1 - \sum_{i=1}^{k-1} P_i - \sum_{i=1}^{k-1} Q_i} \qquad (7.5.16)$$

The variance of the concentration by mass X_k is obtained analogously to the equation (7.4.10):

$$\sigma^2(X_k) = P''_k(1-P''_k) \cdot \frac{g_k}{g}(1 - \sum_{i=1}^{k-1} P_i - \sum_{i=1}^{k-1} Q_i) + P''^2_k \cdot \sigma^2(\sum_{i=1}^{k-1} X_i + \sum_{i=1}^{k-1} Y_i) \qquad (7.5.17)$$

The variance $\sigma^2(\sum_{i=1}^{k-1} X_i + \sum_{i=1}^{k-1} Y_i)$ is given by a recurrence relation which is analogous to equation (7.4.16):

$$\sigma^2(\sum_{i=1}^{k} X_i + \sum_{i=1}^{k} Y_i) = (P_k + Q_k)\left(1 - \frac{P_k + Q_k}{(1 - \sum_{i=1}^{k-1} P_i - \sum_{i=1}^{k-1} Q_i}\right) \cdot \frac{g_k}{g} + $$
$$+ \left(1 - \frac{P_k + Q_k}{1 - \sum_{i=1}^{k-1} P_i - \sum_{i=1}^{k-1} Q_i}\right)^2 \cdot \sigma^2(\sum_{i=1}^{k-1} X_i + \sum_{i=1}^{k-1} Y_i) \qquad (7.5.18)$$

On a transition to continuous distributions we have analogously to equation (7.5.7):

$$\boxed{\sigma^2(R^*) = \frac{(1-R)^2}{g} \int_0^R \frac{g(\xi)}{(1-\xi)^2}\, d\xi} \qquad (7.5.19)$$

The following relations therefore hold:

$$\begin{aligned}
&a) \quad \sum_{i=1}^{k} X_i + \sum_{i=1}^{k} Y_i = R^* \\
&b) \quad \sum_{i=1}^{k} P_i + \sum_{i=1}^{k} Q_i = R \\
&c) \quad X_i + Y_i = d\,R^* \\
&d) \quad P_i + Q_i = d\,R
\end{aligned} \qquad (7.5.20)$$

To calculate the variance of the summation $\sum_{i=1}^{k} X_i$ the following expression is expanded analogously to section (7.4):

$$\sigma^2(\sum_{i=1}^{k} X_k) = \sigma^2(X_k + \sum_{i=1}^{k-1} X_i)$$

$$\sigma^2(\sum_{i=1}^{k} X_k) = \sigma^2(X_k) + \sigma^2(\sum_{i=1}^{k-1} X_i) + 2\,E(X_k \cdot \sum_{i=1}^{k-1} X_i) - 2\,P_k \cdot \sum_{i=1}^{k-1} P_i \qquad (7.5.21)$$

The expectation $E(X_k \cdot \sum_{i=1}^{k-1} X_i)$ is at first still unknown. By means of the rule for calculating conditional expectations we obtain

$$E(X_k \cdot (\sum_{i=1}^{k-1} X_i + \sum_{i=1}^{k-1} Y_i)) = E\left\{ E(X_k \cdot (\sum_{i=1}^{k-1} X_i + \sum_{i=1}^{k-1} Y_i) \,\middle|\, (\sum_{i=1}^{k-1} X_i + \sum_{i=1}^{k-1} Y_i)\right\}$$

$$= E\left\{ (\sum_{i=1}^{k-1} X_i + \sum_{i=1}^{k-1} Y_i) \cdot E(X_k \,\middle|\, (\sum_{i=1}^{k-1} X_i + \sum_{i=1}^{k-1} Y_i)\right\}$$

With

$$E(X_k \,\Big|\, (\sum_{i=1}^{k-1} X_i + \sum_{i=1}^{k-1} Y_i)) = P_k'' \cdot (1 - (\sum_{i=1}^{k-1} X_i + \sum_{i=1}^{k-1} Y_i))$$

it follows that:

$$E(X_k \cdot (\sum_{i=1}^{k-1} X_i + \sum_{i=1}^{k-1} Y_i)) = P_k'' \cdot E(\sum_{i=1}^{k-1} X_i + \sum_{i=1}^{k-1} Y_i) - P_k'' \cdot E(\sum_{i=1}^{k-1} X_i + \sum_{i=1}^{k-1} Y_i)^2$$

$$E(X_k \cdot \sum_{i=1}^{k-1} X_i) + E(X_k \cdot \sum_{i=1}^{k-1} Y_i) = P_k''(\sum_{i=1}^{k-1} P_i + \sum_{i=1}^{k-1} Q_i) - P_k'' \cdot \sigma^2(\sum_{i=1}^{k-1} X_i + \sum_{i=1}^{k-1} Y_i)$$

$$- P_k'' \cdot E^2(\sum_{i=1}^{k-1} X_i + \sum_{i=1}^{k-1} Y_i) \qquad (7.5.22)$$

Since by assumption, the component (X) has the same grain-size distribution as the component (Y), it follows that:

$$\frac{\sum_{i=1}^{k} P_i}{P} = \frac{\sum_{i=1}^{k} Q_i}{Q} \qquad (7.5.23)$$

The expectations $E(X_k \cdot \sum_{i=1}^{k-1} X_i)$ and $E(X_k \cdot \sum_{i=1}^{k-1} Y_i)$ thus differ only by the factor P/Q:

$$E(X_k \cdot \sum_{i=1}^{k-1} X_i) = \frac{P}{Q} \cdot E(X_k \cdot \sum_{i=1}^{k-1} Y_i) \qquad (7.5.24)$$

We then have from equation (7.5.22):

$$E(X_k \cdot \sum_{i=1}^{k-1} X_i) = P_k \cdot \sum_{i=1}^{k-1} P_i - P \cdot P_k'' \cdot \sigma^2(\sum_{i=1}^{k-1} X_i + \sum_{i=1}^{k-1} Y_i)$$

From the equations (7.5.21) and (7.5.17) a recurrence relation is obtained therefrom for $\sigma^2(\sum_{i=1}^{k} X_i)$:

$$\sigma^2(\sum_{i=1}^{k} X_i) = \sigma^2(X_k) + \sigma^2(\sum_{i=1}^{k-1} X_i) - 2\, P \cdot P_k'' \cdot \sigma^2(\sum_{i=1}^{k-1} X_i + \sum_{i=1}^{k-1} Y_i)$$

$$\boxed{\sigma^2(\sum_{i=1}^{k} X_i) = P_k(1-P_k'') \cdot \frac{g_k}{g} + \sigma^2(\sum_{i=1}^{k-1} X_i) + (P_k''-2P) \cdot P_k'' \cdot \sigma^2(\sum_{i=1}^{k-1} X_i + \sum_{i=1}^{k-1} Y_i)}$$

(7.5.25)

Together with the recurrence relation for $\sigma^2(\sum_{i=1}^{k-1} X_i + \sum_{i=1}^{k-1} Y_i)$ the problem has thus been solved in principle. The transition to continuous distributions produces substantial simplifications. Let:

$$\begin{aligned}
&a) \quad X_k = d\,R_x^*\\
&b) \quad P_k = d\,R_x\\
&c) \quad \sum_{i=1}^{k} X_i = R_x^*\\
&d) \quad \sum_{i=1}^{k} P_i = R_x
\end{aligned} \qquad (7.5.26)$$

Equation (7.5.19) is substituted for $\sigma^2(\sum_{i=1}^{k-1} X_i + \sum_{i=1}^{k-1} Y_i)$ and the following differential equation is obtained:

$$\frac{d\,\sigma^2(R_x^*)}{d\,R_x} = \frac{g_k}{g} - \frac{2\,P(1-R)}{g} \int_0^R \frac{g(\xi)}{(1-\xi)^2}\,d\xi$$

and therefrom

$$\sigma^2(R_x^*) = \frac{1}{g} \cdot \int_0^{R_x} g(\eta)\cdot d\eta - \frac{2\,P}{g} \cdot \int_0^{R_x} (1-R) \cdot \int_0^R \frac{g(\xi)}{(1-\xi)^2}\,d\xi$$

For the case of interest, $\sigma^2(X) = \sigma^2(\sum_{i=1}^{k} X_i)$ the limit $R_x = P$ holds, and with $R_x = P \,.\, R$ it follows that:

$$\boxed{\sigma^2(X) = \frac{P}{g} \cdot \int_0^1 g(R)\cdot dR - \frac{2\,P^2}{g} \cdot \int_0^1 (1-R)\cdot \int_0^R \frac{g(\xi)}{(1-\xi)^2}\,d\xi\;dR} \qquad (7.5.27)$$

Special cases

First the limiting case of a mixture from two even-grained fractions, with $g_x = g_y = g_1$, will be considered as an example:

$$\sigma^2(X) = \frac{g_1}{g} P[R]_o^1 - \frac{g_1}{g} \cdot 2P^2 \cdot \int_0^1 (1-R)\left[\frac{1}{1-\xi}\right]_o^R dR$$

$$\sigma^2(X) = \frac{g_1}{g} \cdot P - \frac{g_1}{g} \cdot 2P^2 \int_0^1 R\, dR$$

$$\sigma^2(X) = \frac{g_1}{g}(P - P^2)$$

$$\sigma^2(X) = P \cdot Q \cdot \frac{g_1}{g} \qquad (7.5.28)$$

The limiting case thus agrees with the formulae, derived above, for mixtures from two even-grained fractions. The assumption that a cumulative screenings distribution is a power distribution produces a clear result:

$$Q_3^x(\delta) = Q_3^y(\delta) = Q_3(\delta) = \left(\frac{\delta}{\delta_{max}}\right)^m \qquad (7.5.29)$$

$Q_3(\delta)$ is the distribution of the total mixture and it is thus identical to the grain-size distribution of the individual components. We obtain from (7.5.29):

$$\delta^3 = \delta_{max}^3 \cdot Q_3^{\frac{3}{m}}$$

For constant density and constant shape factors we thus have:

$$g = g_{max} \cdot Q_3^{\frac{3}{m}} = g_{max} \cdot (1-R)^{\frac{3}{m}} \qquad (7.5.30)$$

Hence from (7.5.27):

$$\sigma^2(X) = \frac{g_{max}}{g} \cdot P \cdot \int_0^1 Q_3^{\frac{3}{m}}\, dQ_3 - \frac{g_{max}}{g} \cdot 2P^2 \cdot \int_0^1 Q_3 \cdot \frac{m}{3-m} \cdot \left(1 - Q_3^{\frac{3-m}{m}}\right) dQ_3$$

$$\sigma^2(X) = \frac{g_{max}}{g} P \cdot m \left\{ \frac{1}{3+m} - \frac{2P}{3-m}\left(\frac{1}{2} - \frac{m}{3+m}\right)\right\}$$

$$\sigma^2(X) = \frac{g_{max}}{g} \frac{m}{3+m} P \cdot Q \tag{7.5.31}$$

$m = 1 \qquad \sigma^2(X) = \frac{1}{4} \frac{g_{max}}{g} P \cdot Q$

$m = 2 \qquad \sigma^2(X) = \frac{2}{5} \frac{g_{max}}{g} P \cdot Q$

$m = 3 \qquad \sigma^2(X) = \frac{1}{2} \frac{g_{max}}{g} P \cdot Q$

$m \Rightarrow \infty \qquad \sigma^2(X) = \frac{g_{max}}{g} P \cdot Q$

The limiting case $m \to \infty$, that is to say the case of an infinitely steep distribution, is identical to a mixture which is even-grained with grain size g_{max}.

7.6 Random mixture for suspensions (Table 7.6.1)

As a rule, suspensions are sampled not by means of samples of constant sample mass g, but they are sampled by taking samples of constant sample volume with the aid of given sampling vessels. However, the previous derivations for samples of constant mass hold completely analogously, if the concentration by mass X is replaced by the concentration by volume X_v or if P is replaced by P_v, the concentration by volume in the total, and if the single-grain masses g_i are replaced by the single-grain volumes v_i. This reduces the problem to the case of a random mixture of two components (solid and liquid), already considered in Section 7.5. Since the "grain sizes " of the liquid are always smaller than the grain sizes of the solid, equation (7.5.9) can correspondingly be used:

$$\sigma^2(X_v) = \frac{Q_v^{\,2} . P_v}{V} \cdot \int_0^1 \frac{v(Q_3)}{(Q_v+P_v \cdot Q_3)^2} d\, Q_3 \tag{7.6.1}$$

Table 7.6.1	Sampling from random-mixed suspensions having an even-grained fraction as the solid
1st task:	Comparison of the experimental variance $s^2(X_v)$ of the concentration of the solid by volume X_v with the variance of the random mixture Fixing the sample suspension volume V
1st step	Fixing the concentration of the solid by volume P_v and the single-grain volume v_1 of the solid.
2nd step	Determination of a minimum packing velocity ε_{min} of the solid in the suspension (for example by centrifugal sedimentation)
3rd step	Determination of the sample variance $s^2(X_v)$ of l samples $s^2(X_v) = \frac{1}{l} \sum_{j=1}^{l} (X_{vj}-P_v)^2$ or $s^2(X_v) = \frac{1}{l-1} \sum_{j=1}^{l} (X_{vj} - \overline{X}_v)^2$ and $\overline{X}_v = \frac{1}{l} \sum_{j=1}^{l} X_{vj}$
4th step	Calculation of the variance of the concentration of the solid by volume
(7.6.5)	$\sigma^2(X_v) = P_v(1-P_v-\varepsilon_{min}) \cdot \frac{v_1}{(1-\varepsilon_{min}) \cdot V}$
5th step	Assignment of a prediction level S(e.q. 95%)
6th step	Find the solutions χ_u^2 and χ_o^2 from chi-square distribution $F_f(\chi^2)$ with the degree of freedom f = l - 1
(5.3.8)	$F(\chi_u^2) = \frac{1}{2}(1-S)$ $F(\chi_o^2) = \frac{1}{2}(1+S)$
7th step	The sample variance agrees with the theoretically expected variance at the preassigned prediction level S, if:
(7.1.6)	$\frac{\chi_u^2}{l-1} < \frac{s^2}{\sigma^2} \leqslant \frac{\chi_o^2}{l-1}$

<u>2nd task:</u> What must the size of the sample suspension volume be so that the concentration of the solid by volume X_V has a certain probability S of not deviating by more than a preassigned amount f_p from the nominal concentration?

1st step Find the solution c from the normal distribution*):

(5.2.4)
$$S = 2\,H(c) - 1$$

2nd step The sample volume V follows from the condition required for the confidence interval of a single sample (l = 1), namely $P_V = X_V \pm c \ \sigma/\sqrt{l}$ where $c \cdot \sigma/\sqrt{l} \leq f_p$ and the equation (7.6.5):

$$V \geq \frac{c^2}{f_p^2} \ P_V(1 - P_V - \varepsilon_{min}) \cdot \frac{v_1}{1 - \varepsilon_{min}}$$

*) It is assumed that the distribution is distributed approximately normally.

Q_3 is the cumulative grain-size distribution of the solid and $P_V = 1 - Q_V$ is the concentration of the solid by volume. In the case of an even-grained fraction of the solid we have:

$$\sigma^2(X_V) = \frac{v_1}{V} \ P_V \cdot Q_V \qquad (7.6.2)$$

When the concentration by volume of the solid is small, that is say $Q_V \to 1$, this equation becomes

$$\sigma^2(X_V) = P_V \cdot \frac{v_1}{V} \qquad (7.6.3)$$

This result for suspensions was derived by Pawlowski[18] by using the Poisson distribution.

However, the result (7.6.2) assumes that the components are present in the sample independently of one another. There must therefore be a finite probability that the sample volume consits of just one of the components. Though, in the case of suspensions, this is possible for the liquid component, it is not possible for the solid, for a given sampling shape. The solid content in the sample is highest when the solid component is removed in the form of its densest packing of porosity ε_{min}. In other words, the solid component cannot be independent of the liquid component. Solid and liquid are correlated with one another. Thus, to be able to make statements about the randomness of the mixing process, the term random mixture is given a fresh definition in this case. The mixture of the highest packing density ε_{min} is considered as the 1st component and the remaining liquid is considered as the 2nd component. This makes it possible again that, as assumed, samples are possible which contain only the 1st component. Thus:

$$\sigma^2(X_V^*) = P_V^* (1 - P_V^*) \cdot \frac{v_1^*}{V}$$

The connection with the pure components is given by the following relation-ships:

$$\text{a)} \quad X_V^* = \frac{V}{1-\varepsilon_{min}}$$

$$\text{b)} \quad P_V^* = \frac{P_V}{1-\varepsilon_{min}} \qquad (7.6.4)$$

$$\text{c)} \quad v_1^* = \frac{v_1}{1-\varepsilon_{min}}$$

The variance of the concentration of the solid by volume X_V is thus given by:

$$\sigma^{*2}(X_V) = P_V(1 - P_V - \varepsilon_{min}) \cdot \frac{v_1}{(1 - \varepsilon_{min}) \cdot V} \qquad (7.6.5)$$

$\sigma^{*2}(X_V)$ is only then equal to $\sigma^2(X_V)$ if it is possible for the packing density $\varepsilon_{min} = 0$, that is to say if there is a finite probability that samples consisting of just one of the components occur during sampling. For very low concentrations of the solid by volume, namely $P_V \to 0$, this exception is of no importance and equation (7.6.5) is identical to equation (7.6.2).

In the other extreme case, if the maximum solid concentration is given, namely $P_V = 1 - \varepsilon_{min}$, the variance $\sigma^{*2}(X_V) = 0$, that is to say it is not necessary in this case to specify a probability level, since each sample concentration is of the same size. This mixture is not a stochastic, but a true mixture.

8 Sampling in a Sample Divider

8.1 Numerical concentration in the case of two or more components (Table 8.1.1)

A sample divider is used not only in stratified sampling but also in stage sampling for obtaining a laboratory or analytical sample. Here the unit to be investigated is either repeatedly divided into two parts, as in a riffler, until the sample size has been reached (Figure 8.1.1), or metered directly into the corresponding portion size on a rotary divider.

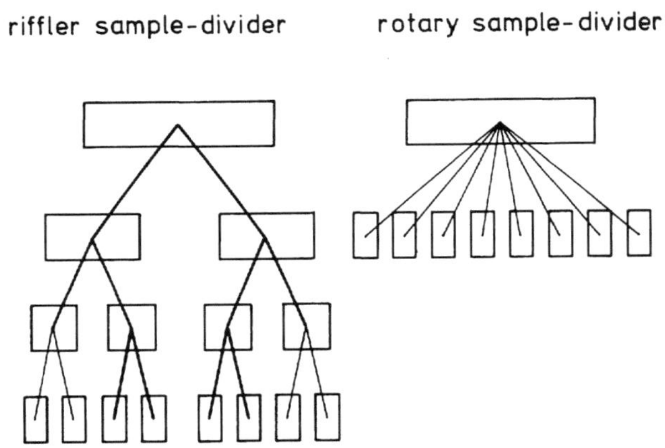

Figure 8.1.1 Diagram of a riffler or rotary sample-divider

In practice, an analysis is normally only carried out on one or, in the case of double sampling, two samples. Whilst in the latter case any two samples can be chosen in the rotary sample-divider, the connection lines in the plan of the riffler-divider must not meet prior to the population. As an example, two such samples and the divisions actually carried out are shown in thick lines in the diagram.

Table 8.1.1 Sampling in a sample divider - numerical concentration x.

1st task: Determination of the expected sampling error:

1st step	Fixing the number N of the particle to be divided.
2nd step	Fixing the number k of the divided final samples.
3rd step	Fixing the nominal numerical concentration p (possibly by estimation by prior experiments $p \cong \bar{x} = \Sigma\, x_i/l$).
4th step	Calculation of the number of particles expected on average in a sample:
(8.1.3)	$E(n) = \frac{N}{k}$
5th step	Calculation of the variance (mean square deviation) of the numerical concentration x

$$\sigma^2(x) \approx \left(1 - \frac{1}{k}\right) \cdot \frac{p(1-p)}{E(n)}$$

2nd task: What is the maximum number of parts into which the material N can be divided, so that there is a certain probability S that the numerical concentration x in a sample does not deviate by more than a preassigned amount f_p from the nominal concentration p?

1st step Find a solution c from the normal distribution*):

$$S = 2\,H(c) - 1 \qquad (5.2.4)$$

2nd step The conditional equation for k follows from the condition required for the confidence interval:

$$k \leq 1 + \frac{f_p^2}{c^2}\,\frac{1}{p(1-p)} \cdot N$$

*)The binomial distribution is approximated by the normal distribution (rule of thumb $n.p > 4$, $n(1-p) > 4$), otherwise c must be determined from the corresponding values of the binomial distribution.

The principle of a sample divider is to divide the starting quantity randomly into the k samples of the final stage. Ideally the components considered are completely independent of one another. It does not therefore matter whether the components are already mixed in the population or whether the completely segregated components are divided one after the other into the k final samples. The concentration distributions in the divided samples therefore cannot permit any inferences about the degree of mixing of the quantity to be divided. The independence of the components during the dividing also has the effect that the sample size varies. Neither the case of a constant number of elements in the sample nor that of constant sample mass thus exists. The derivations from Chapter 7 cannot thus be made use of directly.

It is assumed that the population to be divided consits of N_x elements of the kind (X) and of N_y elements of the kind (Y). These elements are distributed randomly amongst the k final samples. The probability of each element, ending up in one of the samples is to be the same and equal to 1/K. If all the elements are independent of one another, which should be ensured by the design principle of the sample divider, then the probability that two consecutive elements end up in the same sample is $(1/K)^2$. However, the sequence of sub-division does not matter, and thus we have for each component the preconditions required for a binomial distribution. The expectation for the presence, in the individual samples, of n_x elements of the kind (X) or the n_y elements of the kind (Y) is:

$$E(n_x) = \frac{1}{k} \cdot N_x \qquad (8.1.1)$$

$$E(n_y) = \frac{1}{k} \cdot N_y \qquad (8.1.2)$$

and analogously for the sum $n = n_x + n_y$:

$$E(n_x + n_y) = E(n) = \frac{1}{k}(N_x + N_y) = \frac{1}{k} \cdot N \qquad (8.1.3)$$

Correspondingly, the variances are given by:

$$\sigma^2(n_x) = \frac{1}{k}(1 - \frac{1}{k}) \cdot N_x \qquad (8.1.4)$$

$$\sigma^2(n_y) = \frac{1}{k}(1 - \frac{1}{k}) \cdot N_y \qquad (8.1.5)$$

Because n_x and n_y are independent, as defined in the introductory remarks, we have:

$$\sigma^2(n) = \sigma^2(n_x + n_y) = \sigma^2(n_x) + \sigma^2(n_y)$$

$$\sigma^2(n) = \frac{1}{k}(1 - \frac{1}{k})(N_x + N_y) = \frac{1}{k}(1 - \frac{1}{k}) \cdot N \qquad (8.1.6)$$

This result shows that in sample-dividing the sample size n is certainly not constant but, on the contrary, n itself has a binomial distribution with an expectation in accordance with equation (8.1.3) and a variance in accordance with (8.1.6).

In practice, one would like to base inferences about the composition $p = N_x/N$ on the numerical concentration $x = n_x/n$ in the sample. First, the new situation concerning any bias must be checked:

$$E(\frac{n_x}{n}) = E(\frac{n_x}{n_x+n_y})$$

holds approximately with the aid of the two-dimensional Taylor series for independent variables $Cov(n_x, n_y) = 0$:

$$E(g(n_x, n_y)) \approx g(E(n_x), E(n_y)) \qquad (8.1.7)$$

Hence:

$$E(\frac{n_x}{n_x+n_y}) \approx \frac{E(n_x)}{E(n_x) + E(n_y)} = \frac{N_x}{N} = p \qquad (8.1.8)$$

The sample concentration x is thus approximately unbiased. By means of the Gaussian error-propagation law we thus have:

$$\sigma^2(x) = \sigma^2\left(\frac{n_x}{n_x+n_y}\right) \approx \left(\frac{E(n)-E(n_x)}{E^2(n)}\right)^2 \cdot \sigma^2(n_x) + \left(\frac{E(n_x)}{E^2(n)}\right)^2 \cdot \sigma^2(n_y)$$

(8.1.9)

Using the variances from (8.1.4) and (8.1.5):

$$\sigma^2(x) \approx \frac{(1-p)^2}{E^2(n)} \cdot \frac{1}{k}\left(1-\frac{1}{k}\right) \cdot N_x + \frac{p^2}{E^2(n)} \cdot \frac{1}{k}\left(1-\frac{1}{k}\right) \cdot N_y$$

$$\sigma^2(x) \approx \frac{(1-p)^2}{E(n)} \cdot p\left(1-\frac{1}{k}\right) + \frac{p^2}{E(n)}(1-p)\left(1-\frac{1}{k}\right)$$

$$\sigma^2(x) \approx \frac{p(1-p)\left(1-\frac{1}{k}\right)}{E(n)}(1-p+p)$$

$$\boxed{\sigma^2(x) \approx \left(1-\frac{1}{k}\right) \cdot \frac{p(1-p)}{E(n)}} \qquad (8.1.10)$$

If the sample is very small by comparison with the population, that is to say if the population is subdivided into very many units ($k \to \infty$), then (8.1.10) agrees with the result for constant sample size n, if the sample size n is replaced by its expectation.

In the case of a multi-component mixture having the nominal numerical concentrations p_i, it is always possible to group the components in such a manner that a two-component mixture results, since all n_i can be assumed to be independent of one another, and we have in general:

$$\boxed{\sigma^2(x_i) = \left(1-\frac{1}{k}\right) \cdot \frac{p_i(1-p_i)}{E(n)}} \qquad (8.1.11)$$

8.2 Variance of the concentration by mass of two or more components having differing even-grain fractions (Table 8.2.1)

It is also the case in sample division that the determination of the sample mass and of the concentration by mass is much more frequent than the determination of the numerical composition. The mass of the population G divides into the quantities G_x and G_y, and the concentration by mass $G_x/G = P$ and $G_y/G = Q$ are to be estimated by the sample concentration by mass X and Y. The components are to be even-grain fractions having single-grain sizes g_x and g_y. The biasedness of the sample concentration by mass is approximately given by, analogously to equation (8.1.7):

$$E(X) = E\left(\frac{n_x g_x}{n_x \cdot g_x + n_y \cdot g_y}\right) \approx \frac{N_x \cdot g_x}{N_x \cdot g_x + N_y \cdot g_y} = \frac{G_x}{G} = P \qquad (8.2.1)$$

The sample mass g is given by:

$$g = n_x \cdot g_x + n_y \cdot g_y \qquad (8.2.2)$$

and its expectation thus:

$$E(g) = E(n_x) \cdot g_x + E(n_y) \cdot g_y$$

$$E(g) = \frac{N_x}{k} \cdot g_x + \frac{N_y}{k} \cdot g_y = \frac{G}{k} \qquad (8.2.3)$$

The variance of the sample mass is given by:

$$\sigma^2(g) = \sigma^2(n_x \cdot g_x + n_y \cdot g_y) = g_x{}^2 \cdot \sigma^2(n_x) + g_y{}^2 \cdot \sigma^2(n_y)$$

$$\sigma^2(g) = \frac{1}{k}\left(1 - \frac{1}{k}\right)\left\{N_x \cdot g_x{}^2 + N_y \cdot g_y{}^2\right\}$$

$$\boxed{\sigma^2(g) = \left(1 - \frac{1}{k}\right)\left\{P \cdot g_x + Q\, g_y\right\} \cdot E(g)} \qquad (8.2.4)$$

Table 8.2.1 Sampling in a sample divider - concentration by mass.

1st task: Determination of the expected sampling error.

1st step Fixing the total mass G of the material to be divided.

2nd step Fixing the number k of the divided final samples.

3rd step Fixing the nominal concentration by mass P (possible estimation by prior experiments $P \cong \bar{X} = \Sigma \ x_i/l$).

4th step Calculation of the sample mass expected on average.

(8.2.3) $$E(g) = \frac{G}{k}$$

5th step Fixing the single-grain masses g_x and g_y of the components (X) and (Y) or the distributions of the corresponding "average" grain masses:

(8.3.10)
$$g_x \hat{=} \bar{g}_x = \rho_x \cdot k_v^x \cdot M_{3,3}^x \leq g_{x_{max}}$$
$$g_y \hat{=} \bar{g}_y = \rho_y \cdot k_v^y \cdot M_{3,3}^y \leq g_{y_{max}}$$

The maximum single-grain masses g_{xmax}, g_{ymax}, which can be easily found can be used as upper estimates of the sampling error.

6th step Calculation of the variance (mean square deviation) of the concentration by mass:

(8.3.11)
$$\boxed{\sigma^2(X) = \frac{P(1-P)}{E(g)} \left(1 - \frac{1}{k}\right) \left\{(1-P) \cdot g_x + p \cdot g_y\right\}}$$

2nd task: What is the maximum number of parts k into which the material can be divided so that there is a certain probability S that the concentration by mass X in a sample does not deviate by more than a pre-assigned amount f_p from the nominal concentration P?

1st step Find a solution c from the nominal distribution*):

(5.2.4) $$S = 2\,H(c) - 1$$

2nd step A condition equation for k follows from the condition required for the confidence interval:

$$k \leq 1 + \frac{f_p^2}{c^2}\,\frac{G}{P(1-P)\,\{(1-P)\cdot g_x + P\cdot g_y\}}$$

*)The distribution is assumed to be approximately normally distributed.

In sample division the sample mass is also not constant. The variance of the sample concentration by mass must therefore be determined by the error-propagation law:

$$\sigma^2(X) = \sigma^2\left(\frac{n_x\cdot g_x}{n_x\cdot g_x + n_y\cdot g_y}\right) \approx \left(\frac{g_y\cdot E(n_y)}{E^2(g)}\right)^2\cdot\sigma^2(n_x\cdot g_x) + \left(\frac{g_x\cdot E(n_x)}{E^2(g)}\right)^2\cdot\sigma^2(n_y\cdot g_y)$$

(8.2.5)

$$\sigma^2(X) = \left(\frac{g_y\cdot\frac{N_y}{k}}{G^2/k^2}\right)^2\cdot\frac{1}{k}\left(1-\frac{1}{k}\right)\cdot N_x\cdot g_x^{\,2} + \left(\frac{g_x\,\frac{N_x}{k}}{G^2/k^2}\right)^2\cdot\frac{1}{k}\left(1-\frac{1}{k}\right)N_y\cdot g_y^{\,2}$$

$$= \frac{Q^2}{G/k}\cdot\left(1-\frac{1}{k}\right)\cdot\frac{N_x\cdot g_x}{G}\cdot g_x + \frac{P^2}{G/k}\left(1-\frac{1}{k}\right)\frac{N_y\cdot g_y}{G}\cdot g_y$$

$$\sigma^2(X) = \frac{P\cdot Q}{E(g)}\left(1-\frac{1}{k}\right)\left\{Q\cdot g_x + P\cdot g_y\right\} \qquad (8.2.6)$$

For small samples, corresponding to a large k, equation (8.2.6) agrees with the equation of Stange.

In the case of a multi-component mixture having the concentrations by weight $P_i = G_i/G = N_i \cdot g_i/G$ a relationship analogous to the equation (8.2.6) follows:

$$\sigma^2(X_i) = \sigma^2 \left(\frac{n_i \cdot g_i}{n_i \cdot g_i + \sum_{i \neq j} n_j \cdot g_j} \right) \approx \frac{(1-P_i)^2}{E^2(g)} \cdot \sigma^2(n_i \cdot g_i) + \frac{P_i^2}{E^2(g)} \cdot \sigma^2\left(\sum_{i \neq j} n_j \cdot g_j\right) \tag{8.2.7}$$

Because the values n_j are independent, the variance of the sum is equal to the sum of variances:

$$\sigma^2(X_i) = \frac{(1-P_i)^2}{E(g)} \cdot \left(1 - \frac{1}{k}\right) \cdot P_i \cdot g_i + \frac{P_i^2}{E(g)} \cdot \left(1 - \frac{1}{k}\right) \sum_{i \neq j} P_j \cdot g_j$$

$$\boxed{\sigma^2(X_i) = \frac{P_i(1-P_i)}{E(g)} \left(1 - \frac{1}{k}\right) \left\{ (1 - P_i) \cdot g_i + P_i \frac{\sum_{i \neq j} P_j\, g_j}{1 - P_i} \right\}} \tag{8.2.8}$$

In the case of a multi-component mixture, the single-grakn mass g_y is replaced by an average grain mass g_R of the remainder components:

$$\bar{g}_R = \frac{\sum_{i \neq j} P_j \cdot g_j}{1 - P_i} \tag{8.2.9}$$

g_R is not an average in the customary sense; on the contrary it is weighted with the concentration by mass instead of with the usual numerical concentration.

8.3 Variance of the concentrations by number and by mass for a two-material mixture having different grain distributions

If the components to be investigated, (X) and (Y), are in the form of grain distributions having the cumulative sum Q^x and Q^y, this does not affect the variance of the numerical concentration, since the distinction of the components is arbitrary. The individual fractions of the components are collected together and by equation (8.1.10) we have:

$$\sigma^2(\sum_x x_j) \approx (1 - \frac{1}{k}) \cdot \frac{\sum_x p_j(1-\sum_x p_j)}{E(n)} \tag{8.3.1}$$

The index x under the summation sign means that the summation is only to be carried out for the fractions of the component (X). If:

$$\sum_x x_j = x \text{ und } \sum_x p_j = p \tag{8.3.2}$$

then equation (8.3.1) equates with (8.1.10). To calculate the variance of the concentration by mass the total amount is divided similarly into fractions of the component (X) and fractions of the component (Y):

$$\sigma^2(\sum_x X_j) = \sigma^2 \left(\frac{\sum_x n_j g_j}{\sum_x n_j g_j + \sum_y n_j g_j} \right) \tag{8.3.3}$$

Here again the index (x) or (y) denotes the summation over all proportions of the component (X) or component (Y) respectively. An expansion analogous to equations (8.2.7) and (8.2.8) leads to:

$$\sigma^2(\sum_x X_j) = \frac{\sum_x P_j(1-\sum_x P_j)}{E(g)} (1 - \frac{1}{k}) \left\{ (1 - \sum_x P_j) \cdot \frac{\sum_x P_j g_j}{\sum_x P_j} + \right.$$
$$\left. + \sum_x P_j \cdot \frac{\sum_y P_j g_j}{1 - \sum_x P_j} \right\} \tag{8.3.4}$$

X_j and P_j here denote the concentrations by mass of individual fractions of the different components in the total mixture. Fractions of the individual components are collected together and the following abbreviations are introduced:

$$\sum_x X_j = X \text{ und } \sum_x P_j = P \text{ bzw. } \sum_y P_j = Q = 1 - P \qquad (8.3.5)$$

X is the proportion by mass of the component (X) in the sample P, and Q are the corresponding proportions in the population. Equation (8.3.4) can thus be re-written:

$$\sigma^2(X) = \frac{P(1-P)}{E(g)} \left(1 - \frac{1}{k}\right) \left\{ (1-P) \frac{\sum_x P_j g_j}{P} + P \cdot \frac{\sum_y P_j g_j}{Q} \right\} \qquad (8.3.6)$$

The values $\sum_x P_j g_j/P$ and $\sum_x P_j g_j/Q$ describe "average" grain masses of the two components. P_i/P and P_j/Q characterise the proportions by weight of the fractions within these components. From the definitions of particle-size analysis[19] the following simplifying notation can be derived:

$$\sum_x \frac{P_j}{P} g_j = \sum_x \Delta Q^x_{3_j} \cdot g_j = \rho_x k^x_v \sum_x \delta^3 \Delta Q^x_3(\delta) \qquad (8.3.7)$$

k_v^x is a shape factor which connects the geometric volume to the third power of the dispersity value δ, where ρ_x is the density. It has become customary to use the abbreviation $M_{3,3}$ for the sum:

$$M^x_{3,3} = \sum_x \delta^3 \cdot \Delta Q^x_3(\delta) = \sum_x \delta^3 \cdot q^x_3(\delta) : \Delta\delta \qquad (8.3.8)$$

The first index denotes the power of the variable δ, and the second index denotes the dimension of the measure of quantity, here namely the volume or, for a constant density within the components, also the mass.

Analogously, there exists a moment of the component (Y):

$$M^y_{3,3} = \sum_y \delta^3 \Delta Q^y_3(\delta) = \sum_y \delta^3 q^y_3(\delta)\,\Delta\delta \qquad (8.3.9)$$

The "average" grain weights of the components can thus be written as:

$$\text{a)}\quad \bar{g}_x = \rho_x \cdot k^x_v \cdot M^x_{3,3}$$

$$\text{b)}\quad \bar{g}_y = \rho_y \cdot k^y_v \cdot M^y_{3,3} \qquad (8.3.10)$$

$\bar{g}_x$, $\bar{g}_y$ are not averages in the customary sense; on the contrary they are weighted with the mass distribution instead of with the usual numerical distribution.

The variance of the concentration by mass is thus given in a form which is analogous to equation (8.2.6) for two-material mixtures:

$$\boxed{\sigma^2(X) = \frac{P \cdot Q}{E(g)}\,(1 - \frac{1}{k}) \cdot \left\{ Q \cdot \bar{g}_x + P \cdot \bar{g}_y \right\}} \qquad (8.3.11)$$

8.4 Experiments on a rotary sample-divider from Messrs. Retsch

A rotary sample divider from Messrs. Retsch was used to divide mixtures of glass spheres and iron spheres at first into eight vessels and then to subject the content of each vessel again to the same procedure. The 64 samples thus obtained were analysed. The sample masses were first determined and then the iron spheres were separated from the glass spheres with the aid of a magnet and the concentration by mass in the sample was determined.

a) Variations of the sample mass

To investigate the variations of the sample mass two mixtures of 64 g of glass spheres and 64 g of iron spheres respectively were divided. In a further experiment, a mixture of 42.7 g of iron spheres and 85.3 g of glass spheres was divided. (Table 8.4.1).

Table 8.4.1 Variations of the sample mass

P	g_{Gl}	g_{Fe}	g^*	$\sigma^2(g)$	$s^2(g)$
0,5	14 mg	14 mg	14 mg	$2,76 \cdot 10^{-2}\ g^2$	$1,91 \cdot 10^{-2}\ g^2$
0,5	6 mg	6 mg	6 mg	$0,34 \cdot 10^{-2}\ g^2$	$0,35 \cdot 10^{-2}\ g^2$
0,67	5,5 mg	15 mg	8,7mg	$1,71 \cdot 10^{-2}\ g^2$	$1,64 \cdot 10^{-2}\ g^2$

P = Concentration of glass spheres in the total mixture.

g_{Gl} = Single-grain mass of the glass spheres.

g_{Fe} = Single-grain mass of the iron spheres.

g^* = $P \cdot g_{Gl} + (1 - P) \cdot g_{Fe}$

$\sigma^2(g)$ = Variance of the sample mass in accordance with equation (8.2.4) with E(g) = 2 g.

$s^2(g)$ = Sample variance of the sample mass from k = 64 samples.

The experimentally determined variances were compared with the theoretical ones from equation (8.2.4). Assuming the chi-square test to be applicable, all results were within the 95 % confidence interval.

b) Influence of the mean sample mass on the variance of the concentration by mass

In Figure 8.4.1, the sample mass was varied while the mixture composition P = 0.5 with the single-grain masses g_{Gl} = 12.6 mg and g_{Fe} = 15.3 mg was kept constant.

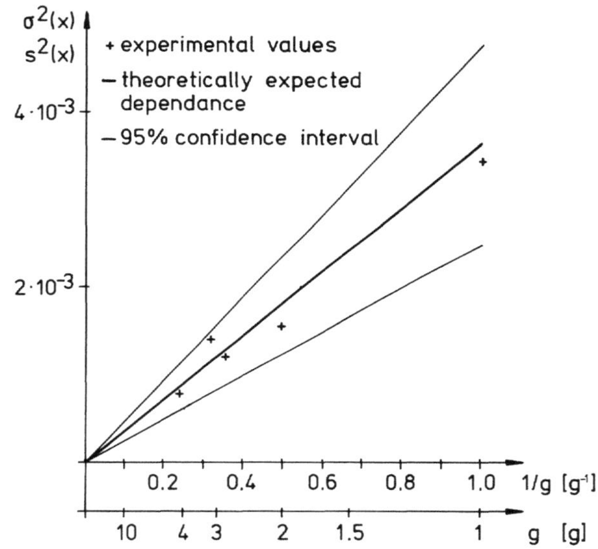

Figure 8.4.1 Influence of the mean sample mass on the variance of the concentration by mass

The solid lines correspond to the variances for the concentration by mass X of the glass spheres, which have been calculated in accordance with equation (8.2.6), together with the 95 % confidence interval of the chi-square distribution. The sample variances from are all within this confidence region.

c) Effect of the mixture composition on the variance of the concentration by mass

Using the same extra components (glass spheres g_{Gl} = 12.6 mg and iron spheres g_{Fe} = 15.3 mg) the composition was varied while keeping constant the mean sample size g = 2 g. The independent variable $P.Q(Q.g_{Gl}+P.g_{Fe})$ was plotted on the abscissa in Figure 8.4.2. Again, the experimental values were all within the 95 % confidence interval.

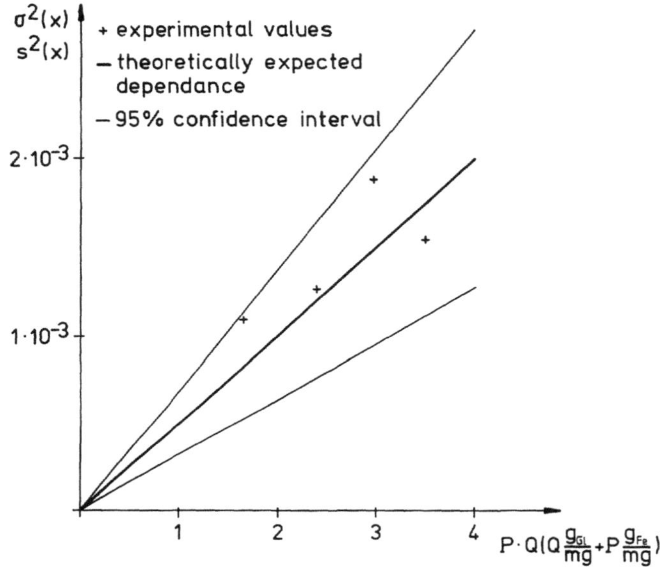

Figure 8.4.2 Effect of mixture composition on the variance of the concentration by mass

9 Sampling for Grain-Size Analysis

9.1 Object and method of sampling

The object of particle-size analysis is to determine the particle-size distribution or a part range, a moment or a mean of the particle-size distribution of a given population[20]. As is the case in any problem of sampling, the population is frequently too large to be investigated completely and a sample has to be taken which can be handled by the means of investigation available in the laboratory, and is then reduced in one or more stages to the size of an analytical sample. Inferences about the population Q must then be based on the value Q* of the distribution measured on the analytical sample. Since the analytical error is usually very small (about 2 %), the sampling error, which alone will be considerd below, should be smaller or, at most, of the same order of size.

Sampling from large quantities of material is the same as sampling from other bulk goods such as ores, fuels and the like, and tools for this purpose are described in Chapter 14. There are two basic methods for taking an analytical sample from a laboratory sample: sample division and sampling from a random mixture.

In sample division, the state of division in the final sample is independent of the degree of mixing, and this type of sampling is therefore advisable for any free-flowing bulk goods, which are always subject to possible segregation phenomena. The riffler divider pours the laboratory sample into several chambers open at the top, from each of which adjacent material flows out at two different sides and is combined into two samples.

Splitting is then continued until the final sample size has been achieved. If the taking of a single sample is envisaged, it is sufficient to divide appropriately just a partial amount.

However, when several samples are to be taken, care has to be taken that the splitting lines only join up at the laboratory sample. In a rotating sample-splitter, the slow-flowing material stream is passed, by means of a vibrating chute, to a central feeding device. There, the stream of material is divided into a number of rotating vessels. In this division, the probability of each particle ending up in any one of the vessels should be the same. If the analytical sample size can be achieved by the splitting, then each sample enjoys equal rights and samples are available at one for double sampling or multiple sampling. However, if the sample has to be divided further, then what was said in connection with the riffler divider is still valid, that is to say the splitting lines of the samples to be investigated must only meet at the starting sample. If no sample splitting device is available, it is possible to make do with the process of coning and quartering, in which the sample is allowed to flow out through a funnel and thus forms a cone. The cone is quartered by means of two cuts vertical to one another. The parts which are diagonally opposite one another are combined to form a part sample and can be further divided if required.

The concept of sampling from a random mixture can usually be realized for free-flowing material only in the case of suspensions. If the particle size and hence the sedimentation velocity are not too large, then this state can be prepared an maintained by stirring and shaking. Sampling from a suspension is of course appropriate when the particle-size analysis is to be carried out in this state. According to Koglin[20)], a method which has proved suitable is first to prepare part amounts of the order of size of 10 g from a laboratory sample of about 1 kg by the processes described above of sample splitting, and to mix these part amounts with about 1 l of suspending fluid and to prepare dispersions. Analytical samples are then taken from the shaken suspension, which is present as a random mixture, by means of a pipette, from a few cm above the bottom of the vessel. In order to prevent particles adhering to the walls of the pipette, it must be rinsed thoroughly.

9.2 Sampling error for sampling with a constant particle number n from a numerical distribution Q_o (Table 9.2.1)

Counting methods for determining the particle-size distribution present the problem of how many particles have to be counted in order to obtain a sampling error fulfilling the condition required, namely being of the same order of size as the analytical error. If $Q_o(\delta)$ is the cumulative sum, namely the number of all particles which are smaller than or equal to the given particle size δ, the total number to be investigated can then be divided into two sub-sets having particle sizes $\xi \leq \delta$. and $\xi > \delta$. The concentrations of these subsets in the total number thus correspond exactly to the cumulative sum $Q_o(\delta)$ or $1 - Q_o(\delta)$, respectively. The variance of the measured cumulative sum $Q_o^*(\delta)$, as the proportion of particles $\xi \leq \delta$ in the sample, is thus given, without taking into account the finite population correction factor, by:

$$\sigma^2(Q_o^*(\delta)) = \frac{Q_o(\delta)\,(1 - Q_o(\delta))}{n} \tag{9.2.1}$$

If the error f_P is then to have the probability P of not being larger than $f_P = c\,\sigma\,(Q_o^*)$, then the number of particles which have to be counted is given by

$$n = \frac{c^2}{f_P^2} \cdot Q_o(\delta)\,(1 - Q_o(\delta)) \tag{9.2.2}$$

If it is now assumed that the number of particles is sufficiently large, then the Q_o^* are approximately normally distributed and Table 9.2.1 gives, for $Q_o = 0.5$, the numbers n required for particles in the sample, so that the probability P = 95 % ($c \cong 1.96$) that various f_P are not exceeded.

Table 9.2.1 Sampling error in grain-size analysis for samples having constant particle number n

1st task — What is the size of the sample error of the measured cumulative numerical distribution $Q_o(\delta)$ and the cumulative volume distribution $Q_3(\delta)$?

1st step — Fixing the sample size n

2nd step — Determination of the grain-size distribution Q_o or, possibly by estimation with the mean from l samples

3rd step — Determination of the mean solid volume E(V) expected and the "average" single-grain volumes by prior experiments

(9.3.2)
$$\bar{v}_{1,3} = \frac{k_v}{Q_3} \int_0^{\delta} \xi^3 q_3(\xi)\, d\xi \le k_v \cdot \delta^3$$
$$\bar{v}_{2,3} = \frac{k_v}{1-Q_3} \int_{\delta}^{\infty} \xi^3 q_3(\xi)\, d\xi \le k_v \cdot \delta_{max}^3$$

$$E(V) = k_v \cdot n \cdot \int_{\delta_{min}}^{\delta_{max}} \xi^3 q_0(\xi) d\xi$$

The maximum single-grain volumes $k_v \delta^3$ or $k_v \delta_{max}^3$ can, in each case, be used as upper estimated of the sampling error.

4th step — Calculation of the sampling error

(9.2.1)
$$\sigma^2(Q_o^*) = \frac{Q_o(1-Q_o)}{n}$$

(9.3.3)
$$\sigma^2(Q_3^*) = \frac{Q_3(1-Q_3)}{E(V)} \cdot \left\{(1-Q_3) \cdot \bar{v}_{1,3} + Q_3 \cdot \bar{v}_{2,3}\right\}$$

2nd task: — What is the maximum number of parts k into which the material to be split can be divided, so that there is a certain probability S that the grain-size distributions in the sample do not deviate by more than a pre-assigned amount f_p from the true distributions Q_o or Q_3?

1st step Find the solution c from the normal distribution*): H(c):

(5.2.4) $S = 2\,H(c) - 1$

2nd step The conditional equation for k follows from the condition required for the confidence interval:

(9.2.2)
$$n \geq \frac{c^2}{f_p^2} Q_o(1 - Q_o)$$
$$E(V) \geq \frac{c^2}{f_p^2} Q_3(1-Q_3) \left\{(1 - Q_3) \cdot \bar{v}_{1,3} + Q_3 \cdot \bar{v}_{2,3}\right\}$$

*)The distribution is assumed to be approximately normally distributed.

<u>Table 9.2.2</u> The number n of particles in the sample, which is required to give the probability P = 95 % that the sampling error f_P is not exceeded.

f_P	0,005	0,01	0,02	0,05	0,1
n	40 000	10 000	2 500	400	100

The cumulative sum $Q_o = 0.5$ was chosen because the function (9.2.1), that is to say the variance $\sigma^2(Q_o*(\delta))$ and hence the error, is maximum at this value.

9.3 Sampling error for sampling with a constant particle number n from a cumulative volume distribution $Q_3(\delta)$ (Table 9.2.1)

Grain-size distributions $Q_3{}^*$, which are determined with the aid of counters, are frequently recalculated, with the aid of assumed or determined grain-shape factores, to give $Q_3{}^*$. In order to find the error in the determination of $Q_3{}^*$, in comparison with the real distribution Q_3, the variance from equation (9.2.1) has to be recalculated. Here, the realization that sampling in the sample divider agrees, for $k \to \infty$, with sampling with a constant number n is used. If the grain distribution to be investigated is divided into a set having a particle size $\xi \leq \delta$ and $\xi \leq \delta$, the problem is quite clearly one of finding the variance from a mixture of two grain distributions. For $1/k \ll 1$, we can hence write down immediately:

$$\sigma^2(Q_3^*(\delta)) = \frac{Q_3(1-Q_3)}{E(V)} \left\{ (1-Q_3) \frac{k_v \int_0^\delta \xi^3 q_3(\xi)\, d\xi}{Q_3} + Q_3 \frac{k_v \int_\delta^\infty \xi^3 q_3(\xi) d\xi}{1-Q_3} \right\} \tag{9.3.1}$$

E(V) is the expectation of the volume of the solids in the sample.

If "average" grain volumes of the fractions to be considered are introduced with the grain-shape factor $k_v = v_k/\xi^3$ namely

a) $$\bar{v}_{1,3} = \frac{k_v}{Q_3} \int_0^\delta \xi^3 q_3(\xi)\, d\xi$$

and (9.3.2)

b) $$\bar{v}_{2,3} = \frac{k_v}{1-Q_3} \int_\delta^\infty \xi^3 q_3(\xi)\, d\xi$$

then we can simplify as follows:

$$\sigma^2(Q_3^*(\delta)) = \frac{Q_3(1-Q_3)}{E(V)} \left\{ (1 - Q_3) \cdot \bar{v}_1 + Q_3 \cdot \bar{v}_2 \right\} \qquad (9.3.3)$$

$\bar{v}_{1,3}$ and $\bar{v}_{2,3}$ are not averages in the customary sense. On the contrary, instead of being weighted with the numerical distribution they are weighted with the volume distribution.

As has been the case for all formulae derived hitherto, the variance to be determined contains the unknown fractions Q_3 and $1 - Q_3$, which thus usually have to be estimated first from measurements. Here, in addition, the "average" grain weights $\bar{v}_{1,3}$ and $\bar{v}_{2,3}$ have to be determined. Because the density distribution q_3 is unknown, they must also be determined from the experimental results themselves. However, frequently, estimated values are sufficient in practice.

Whilst in the case of the numerical distribution the maximum of the variance is always at $Q_o = 0.5$, the maximum in the case of the cumulative volume distribution is dependent on the ratio $\bar{v}_{2,3}/\bar{v}_{1,3}$ and the maximum thus shifts to higher Q_3 values with an increasing ratio, but for $\bar{v}_{2,3}/\bar{v}_{1,3}$ equal to ∞ it tends to the constant value 2/3 (Figure 8.3.1).

However, the absolute value for the maximum of $\sigma^2(Q_3(\delta))$ constantly decreases for a "coarser" grain kept at a constant value, but it also tends relatively rapidly to the constant value $4/27 \cdot \bar{v}_{2,3}/E(V)$ (Figure 9.3.2). This characterizes the dominating influence of coarse grain on the accuracy of the analytical result.

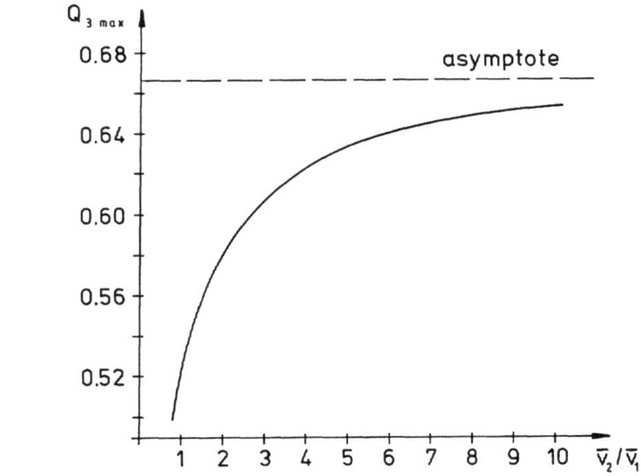

Figure 9.3.1 Q_3 value of maximum variances in dependence on the ratio $\bar{v}_{2,3}/\bar{v}_{1,3}$ for sampling with a constant particle number.

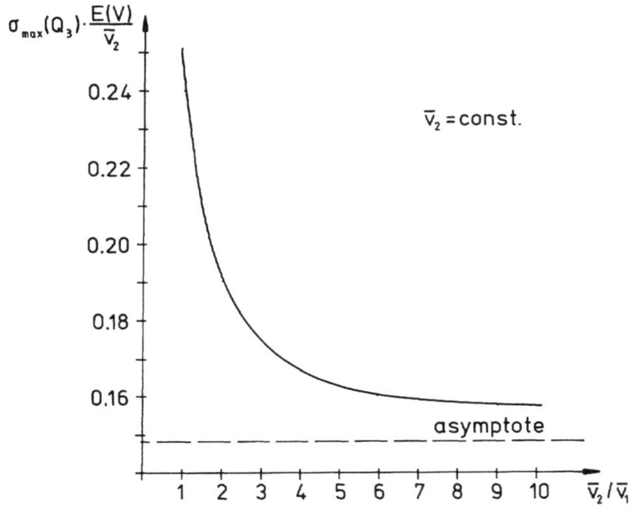

Figure 9.3.2 Maximum variances, of measured Q_3 distributions, in units of $\bar{v}_{2,3}/E(V)$ in dependence on the ratio $\bar{v}_2/\bar{v}_1$ for a constant $\bar{v}_2$

9.4 Sampling error of a cumulative volume distribution $Q_3(\delta)$ for constant sample masses or volumes of solid in the sample (Table 9.4.1)

If samples of constant mass are taken from a population of a particle-size distribution to be investigated, then, for an assumed constant density, the volume of the solid V is also constant and the formulae also hold correspondingly for the volume fractions of the Q_3 distribution. If the discrete fractions are replaced by continuous Q_3 distribution, then the following correspondences hold:

$$
\begin{aligned}
&\text{a)} \quad X_k = -\,d\,Q_3^* \\
&\text{b)} \quad P_k = -\,dQ_3 \\
&\text{c)} \quad \sum_{i=1}^{k} P_i = 1 - Q_3 \\
&\text{d)} \quad \sum_{i=1}^{k} X_i = 1 - Q_3^*
\end{aligned}
\qquad (9.4.1)
$$

If we further assume that dQ_3 is negligibly small compared with 1, then equation (9.4.16) simplifies and we obtain:

$$\sigma^2(Q_3^*(\delta)) - \sigma^2(Q_3^*(\delta) - dQ_3^*(\delta)) \approx -\,dQ_3 \cdot \frac{v_\delta}{V} + \frac{2\,dQ_3}{Q_3} \cdot \sigma^2(Q_3^*) \qquad (9.4.2)$$

v_δ is the grain volume of the dispersity value δ considered. Since $\sigma^2(Q_3^*)$ is a function of the cumulative volume distribution Q_3, equation (9.4.2) is a differential equation for $\sigma^2(Q_3^*(\delta))$, the solution of which yields:

$$\boxed{\sigma^2(Q_3^*(\delta)) = \frac{Q_3^{\,2}}{V} \int_{Q_3}^{1} \frac{v_\xi}{\xi^2}\, d\xi} \qquad (9.4.3)$$

Table 9.4.1 Sampling error in grain-size analysis for samples having constant volumes V of solid in the sample

1st task: What is the size of the sampling error for a measured cumulative volume distribution Q_3*?

1st step Fixing the sample size V

2nd step Determination of the cumulative volume distribution Q_3 and the inverse function for the single-grain volumes $v_{Q3} = v(Q_3)$ by prior experiments

3rd step Calculation of the sampling error (variance):

(9.4.3)
$$\sigma^2(Q_3^*) = \frac{Q_3^2}{V} \int_{Q_3}^{1} \frac{v_\xi}{\xi^2} \, d\xi$$

2nd task: What size does the volume of solid in the sample have to be, so that there is a certain probability S that the volume-size distribution Q_3* does not deviate by more than a pre-assigned amount f_p from the "true" value Q_3?

1st step Find the solultion c from the normal distribution*):

(5.2.4)
$$S = 2\,H(c) - 1$$

2nd step It follows from the condition required for the confidence interval that:

$$V \geq \frac{c^2}{f_p^2} Q_3^2 \int_{Q_3}^{1} \frac{v_\xi}{\xi^2} \, d\xi$$

*)The distribution is assumed to be approximately normally distributed,otherwise the equations (5.3.1) have to be used for the estimation of c.

Special cases

1) For v_δ = constant, that is to say for a mixture of equally large particles, it follows that:

$$\sigma^2(Q_3^*(\delta) = \frac{v_\delta}{V} \cdot Q_3^2 \int_{Q_3}^{1} \frac{1}{\xi^2}\, d\xi = \frac{v_\delta}{V}\, Q_3^2 \left[-\frac{1}{\xi} \right]_{Q_3}^{1}$$

$$\sigma^2(Q_3^*(\delta)) = Q_3(1 - Q_3) \cdot \frac{v_\delta}{V}$$

The variance thus agrees with the formulae derived earlier.

2) If the particle size distribution is approximated by a power distribution and the shape factors are assumed to be constant, then the integral can be evaluated in elementary fashion

$$Q_3(\delta) = \left(\frac{\delta}{\delta_{max}} \right)^m \tag{9.4.4}$$

Integration leads to:

$$\boxed{\sigma^2(Q_3^*(\delta)) = \frac{v_{max}}{V} \, \frac{m}{3-m} \, Q_3^2 \left\{ 1 - Q_3^{\frac{3-m}{m}} \right\}} \tag{9.4.5}$$

In correspondence with the determination of the necessary particle number from Section 9.2, the equations (9.4.3) and (9.4.5) can be used for determining, with a pre-assigned error f_P which shall have the probability P of not being larger than, for example, $f_P = t \cdot \sigma\ (Q_3{}^*)$, the sample sizes V or, for the density ρ, their mass g, necessary for this purpose.

9.5 Sampling error of a numerical distribution $Q_o(\delta)$ or of a cumulative volume distribution $Q_3(\delta)$ in sampling with a constant suspension volume V_{Sus} from a suspension (Table 9.5.1)

Like sampling by splitting (Section 9.6) particle-size analysis in a suspension is one of the most frequently used processes. Particle-analysis instruments as different as photometer, photosedimentometer, sedimentation balance and wet-screening process the product to be investigated in a suspended form. Even dust measurements in air and gases usually use, in the investigation, a given and constant gas space. In all these cases, neither the number of particles to be investigated nor the solid volume (or weight) is constant. For this case, equations (9.4.3) and (9.2.1) are conditional variances with the condition n = constant and V = constant.

Using the rules for calculating conditional variances:

$$\sigma^2(Q_o^*(\delta)) = E_n(\sigma^2(Q_o^*|n)) + \sigma_n^{\,2}(E(Q_o^*|n)) \qquad (9.5.1)$$

and

$$\sigma^2(Q_3^*(\delta)) = E_V(\sigma^2(Q_3^*|V)) + \sigma_V^{\,2}(E(Q_3^*|V)) \qquad (9.5.2)$$

The conditional expectations $E(Q_o{}^*\ n)$ and $E(Q_3{}^*\ V)$ are constant, that is to say the 2nd terms of equations (9.5.1) and (9.5.2) become zero. By means of the approximation formula for expectations the following are calculated:

$$\sigma^2(Q_o^*(\delta)) = E_n\left(\frac{Q_o(1-Q_o)}{n}\right) \approx \frac{Q_o(1-Q_o)}{E(n)}\left(1+\frac{\sigma^2(n)}{E^2(n)}\right) \qquad (9.5.3)$$

$$\sigma^2(Q_3(\delta)) = E_V\left(\frac{Q_3^{\,2}}{V}\int_{Q_3}^{1}\frac{v_\xi}{\xi^2}\,d\xi\right) = \frac{Q_3^{\,2}}{E(V)}\int_{Q_3}^{1}\frac{v_\xi}{\xi^2}d\xi\cdot\left(1+\frac{\sigma^2(V)}{E^2(V)}\right) \qquad (9.5.4)$$

The variances are thus formed analogously with the expectation E(n) instead of n and E(V) instead of V and corrected by the expression in brackets. The solid can be considered as a mixture which has the liquid as a 2nd component. If the mixture is assumed is stochastically homogeneous, then $\sigma^2(n)$ and $\sigma^2(V)$ can be calculated:

$\sigma^2(n)$:

$c_z = E(n)/V_{Susp}$ is defined as the numerical concentration of the particles in the suspension. Liquid particles or, in the case of gas suspensions, infinitely small space elements are considered as particles having $E(n^*)$, and the total number of particles in the suspension volume is given by

$$N = E(n) + E(n^*) \tag{9.5.5}$$

The numerical concentration of the solid particles is thus:

$$c_o = \frac{E(n)}{E(n) + E(n^*)} = \frac{c_z \cdot V_{Susp}}{N} \ll 1 \tag{9.5.6}$$

Because of the large number of liquid particles or space elements in comparison to the number of solid particles, c_o is certainly substantially less than one and it follows that:

$$\sigma^2(n) = c_o(1 - c_o) \cdot N \approx c_o \cdot N = E(n) \tag{9.5.7}$$

$$E(n) = c_z \cdot V_{Susp} \tag{9.5.8}$$

$\sigma^2(V)$:

$c_v = E(V)/V_{Susp}$ is defined analogously as the concentration by volume of the solids. The variance is thus given by:

$$\sigma^2(V) = c_v(1 - c_v) \cdot \bar{v}_{max} \cdot V_{Susp} \tag{9.5.9}$$

Table 9.5.1 Sampling error in the grain-size analysis on a suspension with the sample suspension volume V_{Susp}

1st task: What size is the sampling error for a measured cumulative numerical distribution Q_o* or a cumulative volume distribution Q_3*?

1st step Choice of sample size V_{Susp} and determination of the numerical concentration c_z or the concentration by volume of the solid c_v in the suspension

2nd step Determination of the Q_o- or Q_3-distribution by estimation from prior experiments and hence also determination of the inverse function for the single-grain volumes $v_{Q3} = v(Q_3)$.

3rd step Calculation of the sampling error:

(9.5.12)

$$\sigma^2(Q_o^*) \approx \frac{Q_o(1-Q_o)}{c_z \cdot V_{Susp}} \left(1 + \frac{1}{c_z\, V_{Susp}}\right)$$

$$\sigma^2(Q_3^*) \lesssim \frac{Q_3^2}{c_v \cdot V_{Susp}} \int_{Q_3}^{1} \frac{v_\xi}{\xi^2}\, d\xi \cdot \left(1 + \frac{v_{max}}{c_v \cdot V_{Susp}}\right)$$

2nd task: What is the minimum size of a suspension V_{Susp} so that there is a certain probability S that the numerical distribution Q_3* or the volume-size distribution Q_3* does not deviate by more than a pre-assigned amount f_p from the true values Q_o or Q_3?

1st step Find the solution c from the normal distribution*):

(5.2.4)

$$S = 2\,H(c) - 1$$

*)The distribution is assumed to be approximately normally distributed.

2nd step It follows from the condition required for the confidence interval that:

$$V_{Susp} \geq \frac{c^2}{f_p^2} Q_o(1 - Q_o) \left\{ 1 + \frac{1}{c_z V_{Susp}} \right\}$$

$$V_{Susp} \geq \frac{c^2}{f_p^2 \cdot c_v} Q_3^2 \int_{Q_3}^{1} \frac{v_\xi}{\xi^2} d\xi \cdot (1 + \frac{v_{max}}{c_v \cdot V_{Susp}})$$

The terms $1/(c_z V_{Susp})$ and $v_{max}/c_v \cdot V_{Susp}$ are usually small compared with 1. V_{Susp} can thus be determined by neglecting the terms for the 1st approximation. The 2nd approximation is then corrected with the 1st approximation, etc.

Since c_v is usually very much smaller than one, it follows correspondingly:

$$\sigma^2(V) \approx c_v \cdot v \cdot v_{max} \cdot V_{Susp} \tag{9.5.10}$$

$$E(V) = c_v \cdot V_{Susp} \tag{9.5.11}$$

The corrected formulae for the cumulative numerical distribution Q_o and the cumulative volume distribution Q_3 are thus:

$$\sigma^2(Q_o^*(\delta)) \approx \frac{Q_o(1-Q_o)}{c_z \cdot V_{Susp}} (1 + \frac{1}{c_z V_{Susp}}) \tag{9.5.12}$$

$$\sigma^2(Q_3^*(\delta)) \approx \frac{Q_3^2(\delta)}{c_v V_{Susp}} \int_{Q_3}^{1} \frac{v_\xi}{\xi^2} d\xi \, (1 + \frac{v_{max}}{c_v \cdot V_{Susp}}) \tag{9.5.13}$$

9.6 Sampling error in sample-dividing (Table 9.6.1)

9.6.1 Cumulative numerical distribution $Q_o(\delta)$

The variance for a two-component mixture of differing grain-size distributions was derived in chapter 8.3. In dealing with the error of the cumulative numerical distribution in the analysis of a grain-size distribution, the fractions to be investigated, having particle sizes $\xi \leq \delta$ and $\xi > \delta$, can be considered as a mixture of two components of different grain distribution. If k is the number of samples theoretically expected, then:

$$\sigma^2(Q_o^*(\delta)) = (1 - \frac{1}{k}) \frac{Q_o(1-Q_o)}{E(n)} \qquad (9.6.1)$$

E(n) is the expectation of a number of particles in the sample.

Table 9.6.1 Sampling error in sample-dividing

1st task:	What is the size of the sampling error for measured Q_o*- or Q_3*-distributions?
1st step	Fixing the number of particles E(n) or the volume of solid in the sample E(V) to be expected on average in a sample by the total number of particles N or by the total volume of solid V_{tot} and the number of steps k to be divided: $E(n) = \frac{N}{k}$, $E(V) = \frac{V_{ges}}{k}$
2nd step	Determination of the cumulative numerical distribution Q_o or the cumulative volume distribution Q_3 by estimation from prior experiments
3rd step	Determination of the "average" single-grain volumes $\bar{v}_{1,3} = \frac{k_v}{Q_3} \int_0^{\delta} \xi^3 q_3(\xi)\, d\xi \leq k_v \cdot \delta^3$

$$\bar{v}_{2,3} = \frac{k_v}{1-Q_3} \int_{\delta}^{\infty} \xi^3 q_3(\xi)\, d\xi \le k_v \cdot \delta_{max}^3$$

The more easily determined single-grain volumes k_v i. δ^3 and $k_v\, \delta_{max}^3$ can be used as upper estimates for the sampling error.

4th step Calculation of the sampling error

(9.6.1) $$\sigma^2(Q_o^*) = (1 - \frac{1}{k}) \frac{Q_o(1-Q_o)}{E(n)}$$

(9.6.4) $$\sigma^2(Q_3^*) = (1 - \frac{1}{k}) \frac{Q_3(1-Q_3)}{E(V)} \left\{ (1-Q_3)\cdot \bar{v}_{1,3} + Q_3 \cdot \bar{v}_{2,3} \right\}$$

<u>2nd task:</u> What is the necessary minimum sample size so that the numerical size distribution or the volume size distribution has a certain probability S of not deviating by more than a pre-assigned amount f_p from the "true" values Q_o or Q_3?

1st step Find the solution c from the normal distribution*):

(5.2.4) $$S = 2\,H(c) - 1$$

2nd step It follows from the condition required for the confidence interval that:

$$k \le 1 + \frac{f_p^2}{c^2} \frac{1}{Q_o(1-Q_o)} \cdot N$$

$$k \le 1 + \frac{f_p^2}{c^2} \frac{G_{ges}}{Q_3(1-Q_3)\left\{ (1-Q_3)\cdot \bar{v}_{1,3} + Q_3 \cdot \bar{v}_{2,3} \right\}}$$

*)The equation is assumed to be approximately normally distributed, otherwise the equations (5.1.3) have to be used for the estimation of c.

9.6.2 Cumulative volume distribution $Q_3(\delta)$

Using analogous reasoning to that employed in the case of the cumulative numerical distribution, the cumulative volume distribution (equation 8.3.4) can also be calculated:

$$\sigma^2(Q_3^*(\delta)) = (1-\frac{1}{k})\frac{Q_3(1-Q_3)}{E(V)}\left\{(1-Q_3)\frac{k_v\cdot\int_0^\delta \xi^3 q_3(\xi)\,d\xi}{Q_3} + Q_3\frac{k_v\cdot\int_\delta^\infty \xi^3\cdot q_3(\xi)\,d\xi}{1-Q_3}\right\} \quad (9.6.2)$$

E(V) is the expectation of the volume of the solid in the sample and $k_v = v_\xi/\xi^3$ is the grain-shape factor.

Using the abbreviation

$$\text{a)}\quad \bar{v}_{1,3} \equiv \frac{k_v}{Q_3}\int_0^\delta \xi^3 q_3(\xi)\,d\xi$$

$$\text{b)}\quad \bar{v}_{2,3} \equiv \frac{k_v}{1-Q_3}\int_\delta^\infty \xi^3 q_3(\xi)\,d\xi \quad (9.6.3)$$

we can simplify as follows:

$$\boxed{\sigma^2(Q_3^*(\delta)) = (1-\frac{1}{k})\frac{Q_3(1-Q_3)}{E(V)}\{(1-Q_3)\cdot\bar{v}_{1,3}+Q_3\cdot\bar{v}_{2,3}\}} \quad (9.6.4)$$

$\bar{v}_{1,3}$ and $\bar{v}_{2,3}$ are not averages in the customary sense. On the contrary, they are weighted with the volume distribution rather than the numerical distribution.

Equations (9.6.1) and (9.6.4) agree, except for the factor (1 -1/k), with the corresponding derivations of a constant number.For a large number k, that is to say for the case that the sample is very small by comparison with the population to be divided, the solutions turn into on another. The dependencies on the ratio $\bar{v}_{2,3}/\bar{v}_{1,3}$ which were derived in Section 9.3 hold analogously except for the factor (1-1/k).

10 Sampling Error when Sampling from Ores and Fuels

In the investigation of ores and fuels they are likewise usually present in piece form. However, unlike grain size analysis, the task here is not to determine the proportion of large and small grains; it is rather the case that each grain contains a more or less large content of the target of the investigation, for example metal, moisture or ash. In the text below, the target of the investigation is always assumed to be the metal content of an ore. However, the conclusions can be readily applied to other attributes which are investigated. As in all conventional sampling procedures, samples are taken from the bulk material and frequently mixed and comminuted and finally sub-divided for analysis. Even if the content of metal, moisture or ash is stochastically homogeneous in the starting material, selective comminution[21)] can have the result that the concentration of the material to be investigated differs between the grain classes given. This does not mean that each grain within the same grain class has exactly the same content, but it is to be expected that there will be a distribution of the properties to be investigated in the grain classes also.

The i-th grain class consists of N_i individual grains having a metal content a_{ij}:

a_{ij}:

$$a_{ij} = \frac{G_{ij}^M}{g_i} \tag{10.1}$$

G_{ij}^M is the amount of metal in a grain of the i-th fraction and g_i is the single-grain mass of the i-th fraction. The total metal content of the i-th fraction is thus

$$E(a_{ij}/i) = a_i = \frac{1}{N_i} \sum_{j=1}^{N_i} a_{ij} \tag{10.2}$$

and the variance is given by:

$$\sigma^2(a_{ij}|i) = \sigma_i^2 = \frac{1}{N_i} \sum_{j=1}^{N_i} (a_{ij} - a_i)^2 \tag{10.3}$$

If the total quantity to be investigated has he mass G and if the quantity of the i-th fraction is G_i, then

$$P_i = \frac{G_i}{G} \tag{10.4}$$

is the expected proportion by mass of this fraction.
The proportion of metal G_i^M/G, which the i-th fraction contains in the total amount G and which is to be determined is hence given by:

$$A_i = \frac{G_i^M}{G} = \frac{\sum_{j=1}^{N_i} G_{ij}^M}{G} = \frac{\sum_{j=1}^{N_i} a_{ij} \cdot g_i}{G} = \frac{N_i \cdot g_i \cdot a_i}{G} = \frac{G_i}{G} \cdot a_i = P_i \cdot a_i \tag{10.5}$$

If a sample of size g is taken from the total amount G, then the proportion of the metal g_i^M/g of the i-th fraction in the total sample g is given by:

$$A_i' = \frac{g_i^M}{g} = X_i \cdot a_i' \tag{10.6}$$

We are interested in the variance of this quantity:

$$\sigma^2(A_i') = \sigma^2 \left(\frac{g_i^M}{g} \right) = \sigma^2(X_i \cdot a_i')$$

In order to perform the calculation the conditional variance $\sigma^2(A_i|X_i)$ and the conditional expectation $E(A_i|X_i)$ are determined first. The condition means that it is assumed for the moment that X_i is assumed to be constant in the samples considered.

$$E(A_i'|\,X_i) = E(X_i \cdot a_i'|X_i) = X_i \cdot E(a_i'|\,X_i) = X_i \cdot a_i \qquad (10.7)$$

$$\sigma^2(A_i'\,|X_i) = \sigma^2(X_i \cdot a_i'|X_i) = X_i^2 \cdot \sigma^2(a_i'|X_i) = X_i^{\,2} \cdot \sigma^2(a_i') \qquad (10.8)$$

a_i is equal to the mean of n_i individual particles having metal content a_{ij}:

$$a_i' = \frac{1}{n_i} \sum_{j=1}^{n_i} a_{ij} \qquad (10.9)$$

The variance $\sigma^2(a_i')$ is thus given by:

$$\sigma^2(a_i') = \frac{\sigma^2(a_{ij}\,|\,i)}{n_i} = \frac{\sigma_i^2}{n_i} \qquad (10.10)$$

n_i is linked to the concentration by mass of the sample:

$$n_i = \frac{g}{g_i} \cdot X_i \qquad (10.11)$$

Equation (10.8) hence gives:

$$\sigma^2(A_i'|\,X_i) = X_i \cdot \frac{g_i}{g} \cdot \sigma_i^2 \qquad (10.12)$$

If the condition that X_i be constant is now dropped, it then follows by means of the rule for dealing with conditional variances that:

$$\sigma^2(A_i') = E_{X_i}(\sigma^2(A_i'|X_i)\,) + \sigma^2_{X_i}(E(A_i'|\,X_i) \qquad (10.13)$$

and hence:

$$\boxed{\sigma^2(A_i') = \sigma^2\left(\frac{g_i^M}{g}\right) = P_i \cdot \frac{g_i}{g} \cdot \sigma_i^2 + a_i^{\,2} \cdot \sigma^2(X_i)} \qquad (10.14)$$

The variance of the metal content of an individual fraction in the sample is thus composed of two parts. The critical factor in the first part is the fluctuation of the metal content within the fraction and the second part it is the fluctuation of the fraction proportion X_i. $\sigma^2(X_i)$ can be calcultated according to the type of sampling. If the content a_{ij} of the individual grains is constant, that is $a_{ij} = a_i$, equation (10.14) then simplifies to the immediately understandable result $\sigma^2(A_i) = a_i^2 \cdot \sigma^2(X_i)$. If the proportion X_i is constant in each sample, then A_i' will neverthe-less fluctuate even for constant X_i because of the differing metal contents.

Even more important than the proportion of the metal content which an individual fraction contributes to the total mixture is the problem of finding the total metal content $G_M = \Sigma G_i^M$. If N is the number of fractions present, then the total metal content is given by:

$$A = \sum_{i=1}^{N} \frac{G_i^M}{G} = \sum_{i=1}^{N} P_i \cdot a_i \qquad (10.15)$$

The unbiased metal content of the sample is analogously given by:

$$A' = \sum_{i=1}^{N} \frac{g_i^M}{g} = \sum_{i=1}^{N} X_i \cdot a_i' \qquad (10.16)$$

If the metal quantities g_i^M within the individual fractions are independent of one another the variance of the sum is equal to the sum of the variances:

$$\sigma^2(A') = \sigma^2(\sum_{i=1}^{N} A_i') = \sigma^2(\sum_{i=1}^{N} \frac{g_i^M}{g}) = \sum_{i=1}^{N} \sigma^2(\frac{g_i^M}{g}) = \sum_{i=1}^{N} \sigma^2(A_i')$$

The total metal content in the sample is thus given by:

$$\boxed{\sigma^2(A') = \sum_{i=1}^{N} P_i \cdot \frac{g_i}{g} \cdot \sigma_1^2 + \sum_{i=1}^{N} a_i^2 \cdot \sigma^2(X_i)} \qquad (10.17)$$

The condition of independence of the metal quantities between the fractions is not the case a priori. For example, if all $a_i = a_i' =$ constant A, the the covariance is given by:

$$\mathrm{Cov}\left(\frac{g_i^M}{g}, \frac{g_j^M}{g}\right) = \mathrm{Cov}\,(X_i \cdot a_i', X_j \cdot a_j') = A^2 \cdot \mathrm{Cov}\,(X_i, X_j) \neq 0 \tag{10.18}$$

since the variables X_i and X_j are not independent of one another. The covariances must therefore be taken into account in the case of dependence:

$$\sigma^2(A') = \sum_{i=1}^{N} P_i \cdot \frac{g_i}{g} \cdot \sigma_i^2 + \sum_{i=1}^{N} a_i^2 \cdot \sigma^2(X_i) + \sum_{i \neq j}^{N(N-1)} \mathrm{Cov}\,(a_i' \cdot X_i, a_j' \cdot X_j) \tag{10.19}$$

Example

The example considered is a simple mixture, in a mixing ratio of 1 : 1, of monodisperse metalliferous fraction (1) and a spoils fraction (2). The single-grain masses are related as $g_1/g_2 = 10 : 1$. A sample is to be taken which is 1,000 times the larger single grain. Let the metal content of fraction (1) be $a_1 = 0.75$ and the fluctuations within the layer are $\sigma_1^2(a'_1) = 0.20$. $^2(X_1)$ is calculated from a two-material mixture.

Thus:

$$\sigma^2(X_1) = P_1(1-P_1) \cdot \frac{g_1}{g} = 2{,}5 \cdot 10^{-4}$$

Thus: $P_1 = 0{,}5$, $P_2 = 0{,}5$, $a_1 = 0{,}75$, $a_2 = 0$, $\sigma_1^2 = 0{,}2$, $\sigma_2^2 = 0$, $\sigma^2(X_1) = 2{,}5 \cdot 10^{-4}$.

The expectation of the metal content A is calculated from (10.15) to be

$$E(A') = A = P_1 \cdot a_1 = 0.5 \times 0.75 = 0.375$$

and the variance is given by (10.17) as:

$$\sigma^2(A') = \frac{P_1 \cdot g_1}{g} \cdot \sigma_1^2 + a_1^2 \cdot \sigma^2(X_i) = 1 \cdot 10^{-4} + 1{,}4 \cdot 10^{-4} = 2{,}4 \cdot 10^{-4}$$

11 Investigations of Random Packings

11.1 Definitions and formulation of the problem

In agglomeration and briquetting, in the drying of solids, in filtration, in flows trough a bed - the packing structure, ie. the arrangement of the particles relative to one another, plays a major part in all these examples. In the analysis of such packings an important problem is the determination of the porosity and of the grain-size distribution. The data necessary for the calculation of these values are often obtained by sections through the bed and the section area distributions resulting therefrom. These section area distributions depend of course on the particular state of packing. In the text below it will always be assumed that the packings are random. If it is assumed that all the particles are spheres and so distributed that the average number of midpoints per unit volume is equal to λ , then, in the case of a random packing, these midpoints are poisson-distributed[22]. The spheres forming the packing have a numerical density distribution q(d), where d is the measure dispersity, ie., in the case of spheres, the diameter. Sectioning such a packing density function $z^*(\theta)$ as sections of these spheres. The expectation of this function

$$E(z^*(\theta)) = z(\theta) \tag{11.1.1}$$

can be calculated from the numerical density distribution of the spheres. Wicksell[23] was the first to solve this problem, and we have:

$$z(\theta) = \frac{\theta}{M_{1,0}} \int_{S}^{d_{max}} \frac{q_o(d)\, dd}{\sqrt{(d^2 - \theta^2)}} \tag{11.1.2}$$

$M_{1,0}$ is the first moment of the sphere size distribution:

$$M_{1,0} = \int_{d_{min}}^{d_{max}} d \cdot q_o(d)\, dd \qquad (11.1.3)$$

The lower limit S of integration is either θ or the minimum sphere diameter d_{min}, depending on which of the two diameters is larger. Unless otherwise indicated it is always assumed below that the minimum of the spheres size distribution d_{min} is = 0. Hence, in general, S = θ.

From a parctical point of view, there are 3 main tasks in the present case:

1. It can be shown for packings that the expectation of the area porosity is the same as for the volume porosity. However, an individual section porosity deviates more or less from the expectation. The size of this diviation and hence the error have to be determined.

2. It is assumed in the derivations that the packings are random. They must therefore be checked before application of the equations as to whether this state actually exists. In particular, sphere packings have a tendency to change from a purely stochastic state into an ordered state.

3. As is evident from equation (11.1.2), the distribution of the section circle diameter is in general not identical to the distribution of the sphere diameter. Since z(θ) as the expectation of the section circle area distribution can be determined experimentally, equation (11.1.2) must be solved for $q_o(d)$ in order to determine the sphere diameter distribution. Wicksell[23)] proposed 3 ways of solving this problem:

   a) Determination of all moments $M_{1,0}$ of the sphere diameter distribution from the moments $M_{k,0}$ of the section diameter distribution

b) Solution of the integral equation (11.1.2) and hence obtaining of a direct solution for $q_o(d)$ and

c) Numerical determination of $q_o(d)$ by using histograms.

11.2 Relationship between area porosity and volume porosity

In a thought experiment[24)], an infinitesimally thin slab which has the cross-sectional area F and thickness dh is sectioned from a packing (Figure 11.2.1). In the central plane of this small slab the proportion which the sectional areas of sectioned grains have in the total sectional area is given by:

$$1 - \varepsilon_{F_i} = \frac{f_\theta}{F} \qquad (11.2.1)$$

Since the tickness of the slab in infinitesimally small, the volume porosity within the slab is given by:

$$1 - \varepsilon_{V_i} = \frac{dh \cdot f_\theta}{dh \cdot F} = 1 - \varepsilon_{F_i}$$

and

$$\varepsilon_{V_i} = \varepsilon_{F_i} \qquad (11.2.2)$$

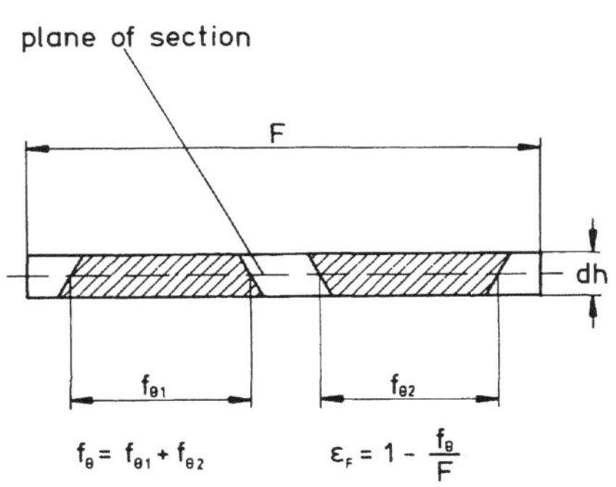

Figure 11.2.1 Diagram showing the principle of the derivation of the expectation of the area porosity

The mean is taken over all possible layers, ie. the expectation is thus given by:

$$E(\varepsilon_{V_i}) = \varepsilon_V = E(\varepsilon_{F_i}) \qquad (11.2.3)$$

ε_{Fi} is thus an unbiased estimate of the volume porosity. This relation is not a consequence of random packing, but it holds for random sections through all types of packing.

11.3 Test for random packing

When a packing is tested for randomness, data such as sphere size distribution and volume porosity are usually not known. However, in order to be able nevertheless to make statements about the randomness of an arrangement, Rumpf[25,26] has suggested considering the distribution of solid and void space in the section as a mixture. If the packing is random, then the sections of solid and the area between the particles must form a random mixture. A packing is sectioned in several planes and the area porosity ε_F is determined. The relationship between the circular areas of the cut solid and the void space is analogous to the case of mixing a grain size distribution with a liquid. The quantity sampled is not the volume but the constant section of the packing F. Equation (7.6.1) for suspensions can thus be applied directly in this case:

$$\sigma^2(1-\varepsilon_F) = \sigma^2(\varepsilon_F) = \frac{E^2(\varepsilon_F)\,(1-E(\varepsilon_F))}{F} \cdot \int_0^1 \frac{f_\theta(Z_2)}{(E(\varepsilon_F)+(1-E(\varepsilon_F)\cdot Z_2)^2}\, dZ_2$$

f_θ is the cross-sectional area of a solid particle; in the case of circular sections it is thus $f_\theta = (\pi/4)\,.\,\theta^2$.

$Z_2 = Z_2(\theta)$ is the area cumulative distribution of the section diameter of the solid particles:

$$Z_2(\theta) = \int_0^\theta z_2(\xi)\,d\xi$$

Thus, for $E(\varepsilon_F) = \varepsilon_V$:

$$\sigma^2(\varepsilon_F) = \frac{\varepsilon_V^2(1-\varepsilon_V)}{F} \int_0^1 \frac{f_\theta(Z_2)}{(\varepsilon_V+(1-\varepsilon_V)Z_2)^2} dZ_2 \qquad (11.3.1)$$

The variance $s^2(\varepsilon_F)$ which is determined experimentally from a finite number of random sections must, in the case of a random packing, agree with the theoretical variance $\sigma^2(\varepsilon_F)$ within preassigned limits with a certain probability. Again it must be stressed here that although in the case of a random packing the theoretical variance $\sigma^2(\varepsilon_F)$ according to equation (11.3.1) is to be expected, on the other hand it is possible that non-random packings produce the same theoretical variance $\sigma^2(\varepsilon_F)$. The inverse of this statement therefore does not hold. In order to prove conclusively that a packing is random (and that the mixture is random) all moments have to be checked, but in practice it is sufficient if the 1st and 2nd moments and hence the variance agree. Equation (11.3.1) at the same time is an answer to the question concerning the confidence probability of the area porosity ε_F as an estimate of the volume porosity ε_V.

11.4 Grain size analysis by determination of the diameters of sectioned solid particles

In heavily agglomerated particles in beds, filter cakes, pellets or sediments, but also in the case of minerals and ores, the individual particles cannot be resolved and are therefore not directly accessible to particle size analysis. In this case, the void spaces are filled if necessary with a hardening agent (eg. Araldite). The packing sample can then be sliced stepwise on a lathe and the section circle diameter distribution $z(\theta)$ or the area porosity ε_F can then be determined on the respective sections. The grain size distribution in the packing is to be inferred from the data obtained.

If the test for random mixture has a positive result, then the link between the density distribution of the section circle diameter and the distribution of the sphere diameter d (equation 11.1.2) can be used to solve this task.

11.4.1 Determination of all moments $M_{n,o}$ of the sphere diameter distribution from the moments $M'_{k,o}$ of the section circle diameter distribution

A distribution is characterized unambiguously if all its moments are known. It is thus sufficient to find a relation which links the moments $M_{n,o}$ with the moments $M'_{k,o}$.

We have:

$$\text{a)}\quad M_{n,o} = \int_0^{d_{max}} d^n \cdot q_o(d)\, dd \qquad (11.4.1)$$

$$\text{b)}\quad M'_{k,o} = \int_0^{\theta_{max}} \theta^k \cdot z(\theta)\, d\theta$$

Together with the equation (11.1.2) and $\theta_{max} = d_{max}$ it follows therefrom that

$$M'_{k,o} = \int_0^{d_{max}} \theta^k\, z(\theta)\, d\theta = \frac{1}{M_{1,0}} \int_0^{d_{max}} \theta^{k+1} \int_\theta^{d_{max}} \frac{q_o(d)\, dd}{\sqrt{(d^2 - \theta^2)}}\, d\theta \qquad (11.4.2)$$

Integration has the result[26]:

$$\text{a)}\quad M'_{2k,o} = \frac{M_{2k+1,0}}{M_{1,0}} \cdot \frac{2\cdot 4\cdot 6\ldots 2\,k}{3\cdot 5\cdot 7\ldots(2\,k+1)}$$

$$(11.4.3)$$

$$\text{b)}\quad M'_{2k-1,0} = \frac{M_{2k,o}}{M_{1,0}} \cdot \frac{1\cdot 3\cdot 5\ldots(2k-1)}{2\cdot 4\cdot 6\ldots 2\,k}$$

Hence the following relations are obtained:

$$
\begin{aligned}
&\text{a)} \quad M_{2,0} = \frac{4}{\pi} M'_{1,0} \cdot M_{1,0} \\
&\text{b)} \quad M_{3,0} = \frac{3}{2} M'_{2,0} \cdot M_{1,0} \\
&\text{c)} \quad M_{4,0} = \frac{16}{3\pi} M'_{3,0} \cdot M_{1,0} \\
&\text{d)} \quad M_{5,0} = \frac{15}{8} M'_{4,0} \cdot M_{1,0}
\end{aligned}
\qquad (11.4.4)
$$

The connection between the first moment $M_{1,0}$ and the moments of the section circle diameter distribution follows from:

$$M'_{-1,0} = \int_0^{d_{max}} \frac{z(\theta)\, d\theta}{\theta} = \frac{1}{M_{1,0}} \int_0^{d_{max}} \int_\theta^{d_{max}} \frac{q_o(d)\, dd}{\sqrt{(d^2 - \theta^2)}}\, d\theta$$

$$M'_{-1,0} = \frac{1}{M_{1,0}} \int_0^{d_{max}} q_o(d)\, dd \cdot \int_0^{d} \frac{1}{\sqrt{d^2 - \theta^2}}\, d\theta$$

$$M'_{-1,0} = \frac{1}{M_{1,0}} \cdot \frac{\pi}{2}$$

$$\boxed{M_{1,0} = \frac{\pi}{2} \frac{1}{M'_{-1,0}}} \qquad (11.4.5)$$

All the moments of the sphere diameter distribution can hence be calculated from the moment of the section circle diameter distribution.

Example

It is assumed that the section circle diameter distribution can be represented by a power function. It would have to be checked here whether there is in fact a sphere diameter distribution which produces a power function. However, for moments which are not too large the power distribution can be used as an aproximation function.

$$Z(\theta) = \left(\frac{\theta}{d_{max}}\right)^m \text{ bzw. } z(\theta) = \frac{m}{d_{max}^m} \cdot \theta^{m-1} \qquad (11.4.6)$$

The moments are then given by:

$$M'_{k,o} = \frac{m}{d_{max}^m} \int_0^{d_{max}} \theta^{k+m-1} \, d\theta$$

$$\boxed{M'_{k,o} = \frac{m}{m+k} \cdot d_{max}^k} \qquad (11.4.7)$$

The expectation of the sphere diameter $E(d) = M_{1,0}$ hence follows from equation (11.4.5):

$$\boxed{E(d) = M_{1,0} = \frac{\pi}{2} \, \frac{m-1}{m} \, d_{max}} \qquad m \geq 1 \qquad (11.4.8)$$

For m = 1, ie. a uniform distribution of the section circle diameter, E(d) = 0, ie. it is not possible that a random packing has an even distribution for the section circle diameter.

Since the expectation E(d) has to be smaller than d_{max}, it must be true that $(\pi/2)(m-1)/m < 1$ and therefrom $m < 1/(1-2/\pi) \cong 2.8$. If m differs only a little from 2.8, the power function is not suitable for estimating the expectation.

The variance of the sphere diameters $\sigma^2(d) = M_{2,0} - M_{1,0}^2$ in the case of a power distribution is given by:

$$\sigma^2(d) = M_{2,0} - M_{1,0}^2 = \frac{4}{\pi} \cdot M'_{1,0} \cdot M_{1,0} - M_{1,0}^2$$

$$\sigma^2(d) = \frac{4}{\pi} \cdot \frac{m}{m+1} \cdot d_{max} \cdot \frac{\pi}{2} \frac{m-1}{m} d_{max} - \left(\frac{\pi}{2} \frac{m-1}{m} d_{max}\right)^2$$

$$\boxed{\sigma^2(d) = d_{max}^2 \left\{ 2 \frac{m-1}{m+1} - \frac{\pi^2}{4} \frac{(m-1)^2}{m^2} \right\}} \qquad (11.4.9)$$

Since the variance $\sigma^2(d)$ must always be larger than 0, the following inequality holds:

$$2 \frac{m-1}{m+1} - \frac{\pi^2}{4} \frac{(m-1)^2}{m^2} \geq 0$$

and hence:

$$m \leq \sqrt{\frac{\pi^2/8}{\pi^2/8 - 1}} \approx 2,3$$

The limiting case of the equality sign would require an even-grain fraction. This contradicts equation (11.4.8) for which, with $m = 2.3$: $E(d) = 0.89\ d_{max}$. If m is only a little smaller than 2.3, the power function is not suitable for estimating the variance $\sigma^2(d)$. Correspondingly it can be shown[27] that in the case of a packing of equal size spheres a power function does not result for the section circle diameter distribution (Figure 11.4.1), but that we have instead:

$$\boxed{z(\theta) = \frac{\theta}{d\sqrt{d^2 - \theta^2}}} \qquad (11.4.10)$$

$$\boxed{Z(\theta) = \int_0^\theta \frac{\xi}{d\sqrt{d^2 - \xi^2}}\, d\xi = 1 - \sqrt{1 - \left(\frac{\theta}{d}\right)^2}}$$

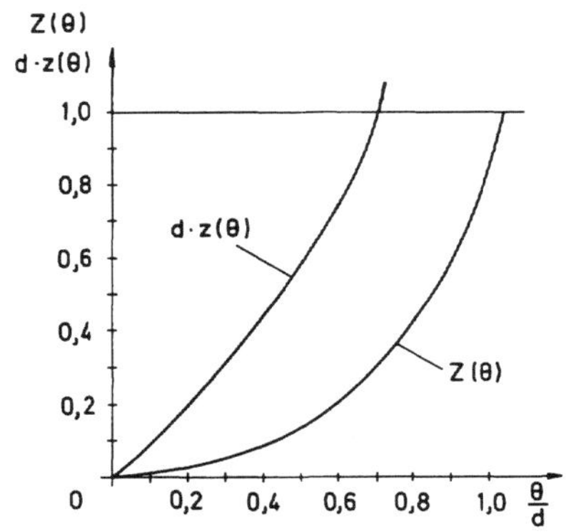

Figure 11.4.1 Section circle distribution of a packing of identical spheres

11.4.2 Solution of the integral equation

The equation (11.1.2) is an integral equation for the unknown distribution $q_o(d)$, the solution of which by Courant/Hilbert[28] is:

$$q_o(d) = -\frac{2\,M_{1,0}}{\pi} \cdot d \cdot \int_{d}^{d_{max}} \frac{\frac{d}{d\theta}\left(\frac{z(\theta)}{\theta}\right)}{\sqrt{\theta^2 - d^2}}\, d\theta \qquad (11.4.11)$$

How this equation is applied is illustrated in the Example below. In this Example, functions $z(\theta)$ are chosen which permit an elementary evaluation of the integral (11.4.11). In the practical real life case, the evaluation must usually be carried out numerically. One possible method of numerical evaluation is carried out at the end of this section with the aid of a histogram. It is assumed in the Example which follows that an analysis of the section circle diameteres produced the numerical cumulative distribution or numerical density distribution shown in Figure (11.4.2).

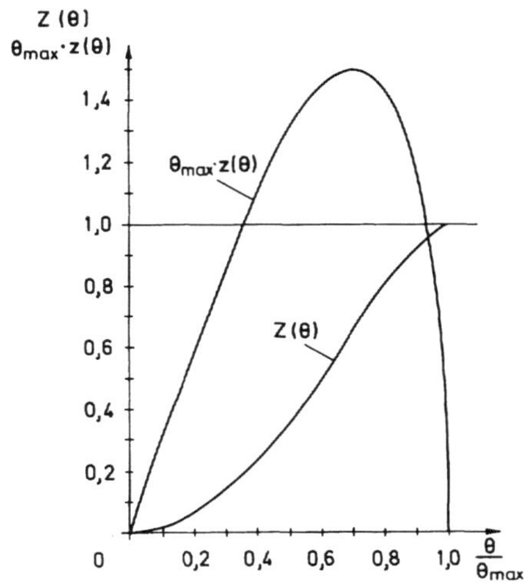

Figure 11.4.2 Section circle distribution of a random sphere packing

The curves can be approximated by the following functions:

a) $$Z(\theta) = 1 - \left(1 - \left(\frac{\theta}{\theta_{max}}\right)^2\right)^{3/2}$$

(11.4.12)

b) $$z(\theta) = \frac{3\,\theta}{\theta_{max}^{3}} \sqrt{\theta_{max}^{2} - \theta^{2}}$$

First the expectation $E(d) = M_{1,0}$ is evaluated by means of equation (11.4.5) to give:

$$M'_{-1,0} = \int_0^{\theta_{max}} \frac{3}{\theta_{max}^{3}} \sqrt{\theta_{max}^{2} - \theta^{2}}\, d\theta = \frac{3}{\theta_{max}^{3}} \left[\frac{\theta}{2} \sqrt{\theta_{max}^{2} - \theta^{2}} + \frac{\theta_{max}^{2}}{2} \arcsin \frac{\theta}{\theta_{max}} \right]_0^{\theta_{max}}$$

$$M'_{-1,0} = \frac{3}{4} \; \frac{\pi}{\theta_{max}} \qquad (11.4.13)$$

and:

$$M_{1,0} = \frac{2}{3}\,\theta_{max} \qquad (11.4.14)$$

Hence the numerical density distribution $q_o(d)$ for $\theta_{max} = d_{max}$ is given by:

$$q_o(d) = -\frac{4\,d_{max}}{3\,\pi}\,d \int_d^{d_{max}} \frac{\frac{d}{d\theta}\left(\frac{3}{d^3_{max}}\sqrt{Q^2_{max} - \theta^2}\right)}{\sqrt{\theta^2 - d^2}}\,d\theta$$

$$q_o(d) = \frac{4}{\pi\,d^2_{max}} \cdot d \cdot \int_d^{d_{max}} \frac{\theta}{\sqrt{d^2_{max} - \theta^2}\sqrt{\theta^2 - d^2}}\,d\theta$$

$$\boxed{q_o(d) = \frac{2}{d^2_{max}} \cdot d} \qquad (11.4.15)$$

The numerical cumulative distribution of the sphere diameter d

$$\boxed{Q_o(d) = \int_0^d \frac{2\,\xi}{d^2_{max}}\,d\xi = \left(\frac{d}{d_{max}}\right)^2} \qquad (11.4.16)$$

is a power distribution with m = 2. It is shown in Figure (11.4.3) in comparison with the numerical cumulative distribution of the section circle diameters. The distribution $Z(\theta)$ is shifted to the finer end as was to be expected. The experimentally determined section circle distribution $z^*(\theta)$ is thus in general not an estimate for the sphere diameter distribution $q_o(d)$. However, there is one exception: if, for example, the expectation of a section circle diameter distribution is given by:

$$z(\theta) = \frac{\theta}{a^2} \cdot e^{-\frac{\theta^2}{2a^2}} \qquad (11.4.17)$$

this then produces for the moment $M'_{-1,0}$:

$$M'_{-1,0} = \frac{1}{a^2} \int_0^{\infty} e^{-\frac{\theta^2}{2a^2}} d\theta = \frac{\sqrt{\pi/2}}{a}$$

and for the expectation of the sphere diameter $M_{1,0}$ with equation (11.4.5):

$$M_{1,0} = a\sqrt{\pi/2}$$

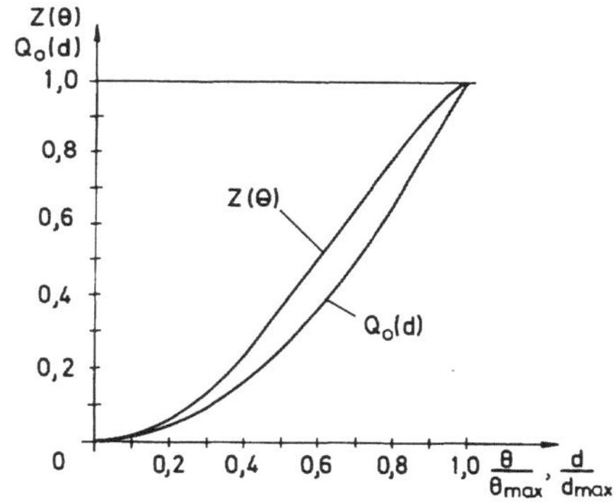

Figure 11.4.3 Comparison between a section circle distribution $Z(\theta)$ and the sphere diameter distribution $Q_o(d)$ which produces it

The numerical density distribution of the sphere diameters is then given by equation (11.4.11):

$$q_o(d) = -\frac{d}{a\sqrt{\pi/2}} \int_d^{\infty} \frac{\frac{d}{d\theta} e^{-\frac{\theta^2}{2a^2}}}{\sqrt{\theta^2 - d^2}} d\theta$$

$$q_o(d) = \frac{d}{a^2\sqrt{\pi/2}} \int_d^{\infty} \frac{\theta \cdot e^{-\frac{\theta^2}{2a^2}}}{\sqrt{\theta^2 - d^2}} d\theta$$

The substitution $\theta^2 - d^2 = u^2$ produces:

$$q_o(d) = \frac{d}{a^2\sqrt{\pi/2}} \int_d^\infty e^{-\frac{u^2+d^2}{2a^2}} du$$

$$q_o(d) = \frac{d}{a^2\sqrt{\pi/2}} \cdot e^{-\frac{d^2}{2a^2}} \int_d^\infty e^{-\frac{u^2}{2a^2}} du$$

$$\boxed{q_o(d) = \frac{d}{a^2} \cdot e^{-\frac{d^2}{2a^2}}} \qquad (11.4.18)$$

In the special case shown here, and only in this case, the distribution of the section circle diameters (equation 11.4.17) agrees with the distribution of the sphere diameters $q_o(d)$. The cumulative distribution is given by:

$$\boxed{Q_o(d) = 1 - e^{-\frac{d^2}{2a^2}}} \qquad (11.4.19)$$

The distributions $q_o(d)$ or $Q_o(d)$ and hence also the section circle diameter distributions $z(\theta)$ or $Z(\theta)$ are shown in Figure 11.4.4.

The expectation $E(d)$ of the distribution has already been calculated:

$$E(d) = M_{1,0} = M'_{1,0} = a\sqrt{\pi/2} \approx 1,25\ a \qquad (11.4.20)$$

The variance follows from:

$$\sigma^2(d) = M_{2,0} - M_{1,0}^2$$

However, with equation (11.4.4a)

$$M_{2,0} = \frac{4}{\pi} \cdot M'_{1,0} \cdot M_{1,0} = \frac{4}{\pi} \cdot M_{1,0}^2$$

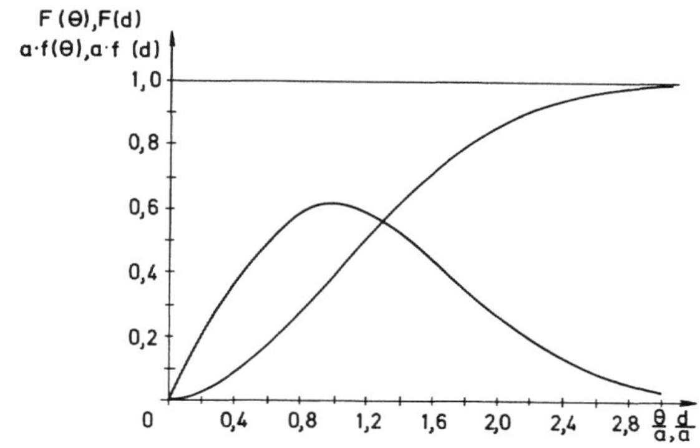

Figure 11.4.4 Special case in which the sphere diameter distribution $Q_o(d)$, $q_o(d)$ and the section circle diameter distribution $Z(\theta)$, $z(\theta)$ agree

and hence:

$$\sigma^2(d) = M_{1,0}^2 \cdot \left(\frac{4}{\pi} - 1\right) = a^2 \cdot \pi/2 \left(\frac{4}{\pi} - 1\right)$$

$$\sigma^2(d) = a^2 (2 - \pi/2) \approx 0,43\ a^2 \qquad (11.4.21)$$

The coefficient of variation is thus:

$$\frac{\sigma(d)}{E(d)} \approx 0,52$$

Further evaluations of the integral equation (11.4.11) are given by Mihira et al.[29]. They deal with the sectinal area distributions for 5 different, given sphere diameter distributions and compare the theoretical results with their own experiments using polyvinyl alcohol foams.

11.4.3 Determination of sphere diameter distributions $q_o(d)$ with the aid of a histogram of section circle diameters

The determination of sphere diameter distributions from a histogram of section circle diameters is very useful in particular if the distribution of the section circle diameters is already given in the form of a histogram because of the experimental method employed. In theory, the range of the section circle diameters extends to zero. Because of the limited resolution of the usually optical measurement method, practical evaluations also have a lower limit. The transformation of the histogram $z(\theta)$ into a histogram $q_o(d)$ can be represented in the form of a matrix equation according to Kausch von Schmeling and Tschoegl[30]. A histogram is the representation of a distribution function as a uniform distribution within certain ranges. The transformation behavior of a uniform distribution of spheres is therefore investigated first with the aid of equation (11.1.2). If the upper (d_1) and lower (d_2) limits of the uniform distribution are known, then a standardized sphere diameter distribution is given by

$$q_o(d) = \frac{1}{d_1 - d_2} \qquad d_2 \le d \le d_1 \tag{11.4.22}$$

This produces, together with equation (11.1.2), a section circle diameter distribution:

$$z(\theta) = \frac{\theta}{M_{1,0}} \int_S^{d_1} \frac{dd}{(d_1 - d_2)\sqrt{d^2 - \theta^2}} \tag{11.4.23}$$

The lower limit S of the integration is either θ or the minimum sphere diameter d_2, depending on which of the two diameters is the larger.

The 1st moment of the uniform distribution is $M_{1,0} = (d_1+d_2)/2$. Integration thus produces:

$$z(\theta) = \frac{2\,\theta}{d_1^{\,2}-d_2^{\,2}} \left(\cosh^{-1}\frac{d_1}{\theta} - \cosh^{-1}\frac{S}{\theta}\right) \qquad (11.4.24)$$

This density distribution of section circle diameters has been plotted in Figure 11.4.5 for $d_2 = d_1/\sqrt{2}$. For small diameters θ, the distribution can be approximated by an expansion into a series:

$$z(\theta) \approx \frac{2\,\theta}{d_1^{\,2}-d_2^{\,2}} \ln\frac{d_1}{d_2} \quad \text{für} \quad \theta \to 0 \qquad (11.4.25)$$

If the theoretically calculated continuous density distribution $n(\theta)$ is divided into, for example, part intervals the size of which decreases in a geometric progression with the factor $\sqrt{2}$, the areas under the curve section of the particular interval give the proportion of this fraction in the total number of all sectioned spheres (Table 11.4.1). In the representation of grain size analyses in the form of a histogram $\sqrt{2}$ produces a gradation which is customarily produced by DIN standards on mesh widths.

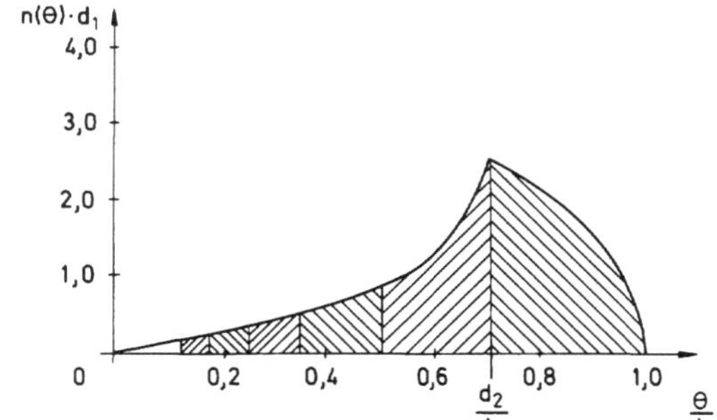

Figure 11.4.5 Section circle diameter of a uniform sphere diameter distribution

Table 11.4.1 Proportion of section circle diameters in a given interval for a uniform distribution of sphere diameters

Interval	Upper limit $\frac{\theta_j}{d_1}$	Proportion $z_j = \int_{\theta_{j+1}}^{\theta_j} z(\theta)\,d\theta$	Proportion of remainder $\sum_{\theta_{j+1}}^{d_1} z_j = \int_{\theta_{j+1}}^{d_1} z(\theta)\,d\theta$
1	1	0,533	0,533
2	$(2)^{-1/2}=0,707$	0,274	0,807
3	$(2)^{-1} = 0,500$	0,102	0,909
4	$(2)^{-3/2}=0,354$	0,047	0,956
5	$(2)^{-4/2}=0,250$	0,022	0,978
6	$(2)^{-5/2}=0,177$	0,011	0,989
7	$(2)^{-6/2}=0,125$	0,0055	0,9945

It can be seen from the last row of Table 11.4.1 that 99.45 % of all sectioned spheres are contained in the first seven intervals.

After the section circle distribution of a uniform distribution of sphere diameters is known, the relationship for a histogram as the sum of various uniform distributions can be derived. If it is assumed that the proportion of spheres in the k-th interval is $q_k = \int q_o(d)dd$, these spheres which are uniformly distributed in the k-th interval produce on sectioning, according to Table 11.4.1, a proportion of section circle diameters in the k-th interval of 0.533 q_k, in the (k+1)-th interval a proportion of 0.274 q_k and so on. For intervals having larger section circle diameters, this fraction does not of course give any proportion. If the sphere diameter distribution consists of many "rectangular distributions", then individual intervals contribute to the histogram of the section circle distribution having identical or smaller diameters. The number within the k-th interval of the section circle diameter is thus given by:

$$z_k = 0,533\, q_k + 0,274\, q_{k-1} + 0,102\, q_{k-2} + 0,047\, q_{k-3} \cdots \qquad (11.4.26)$$

This equation can be written in the form of a matrix equation:

$$z_j = M_{ji} \cdot q_i \qquad (11.4.27)$$

or in the full form:

$$\begin{vmatrix} z_1 \\ z_2 \\ z_3 \\ z_4 \\ \cdot \\ \cdot \\ \cdot \\ z_m \\ \vdots \\ z_{m+1} \\ \vdots \\ z_n \end{vmatrix} = \begin{Vmatrix} 0,533 & 0 & 0 & 0 & & \\ 0,274 & 0,533 & 0 & 0 & & \\ 0,102 & 0,274 & 0,533 & & & \\ 0,047 & 0,102 & 0,274 & 0,533 & & \\ & & & & & \\ & & & & & \\ & & & & & \\ & & & & 0,274 & 0,533 \\ & & & & & \\ & & & & 0,102 & 0,274 \\ & & & & & \\ & & & & & \end{Vmatrix} \begin{vmatrix} q_1 \\ q_2 \\ q_3 \\ q_4 \\ \cdot \\ \cdot \\ \cdot \\ q_{m-1} \\ \\ q_m \\ \\ \\ \end{vmatrix}$$

The actual task was: calculation of the sphere diameter distribution from the section circle distribution. It is given directly by the inversion of equation (11.4.27):

$$q_i = M_{ji}^{-1} \cdot z_j \qquad (11.4.28)$$

M_{ji}^{-1} is the inverse of the matrix M_{ji}. Hence for the matrix indicated above.

$$\begin{vmatrix} q_1 \\ q_2 \\ q_3 \\ q_4 \\ q_5 \\ q_6 \\ q_7 \\ \vdots \\ q_m \end{vmatrix} = \begin{vmatrix} 1,8767 & 0 & 0 & 0 & 0 & \cdots \\ -0,9662 & 1,8767 & 0 & 0 & 0 & \cdots \\ +0,1386 & -0,9662 & 1,8767 & 0 & 0 & \cdots \\ -0,0508 & +0,1386 & -0,9662 & 1,8767 & 0 & \cdots \\ +0,0052 & -0,0508 & +0,1386 & -0,9662 & 1,8767 & \cdots \\ -0,0032 & +0,0052 & -0,0508 & +0,1386 & -0,9662 & \cdots \\ +0,0000 & +0,0032 & +0,0052 & -0,0508 & +0,1386 & \cdots \\ & & & & & \\ & & & & & \end{vmatrix} \begin{vmatrix} z_1 \\ z_2 \\ z_3 \\ z_4 \\ z_5 \\ z_6 \\ z_7 \\ \vdots \\ z_m \\ \vdots \\ z_n \end{vmatrix}$$

12 Sampling from Non-Random Mixtures

12.1 Systematic variance

In a mixer in which the components are completely separate at time $t = t_o$, the spatial concentration distribution will level out in the course of time ($t > t_o$). The mixing process is complete when the same concentration P is to be expected in the whole mixing space, independently of position. Repeating this mixing experiment under identical initial conditions will produce in the long run a highly characteristic concentration profile $P(r,t)$ at time t (Figure 12.1.1).

In the one-dimensional mixing process assumed here the mixing space is divided into segments of size ΔL in which the mean concentration can be assumed to be constant and equal to $P(r,t)$. On taking samples which have a sample mass g the measured concentration $X(r,t)$ will fluctuate randomly about the concentration $P(r,t)$ expected for the particular segment. If the same number of samples of equal size are taken from the

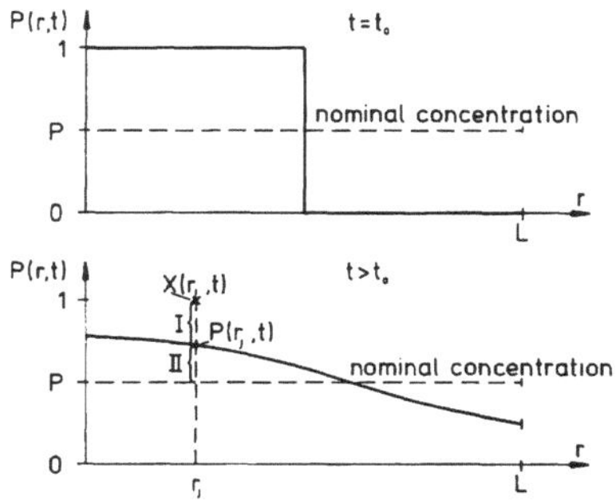

Figure 12.1.1 Concentration distribution $P(r,t)$ at time $t = t_o$ and at time $t > t_o$, where L denotes the length of the mixing drum. I random deviation from $P(r,t)$, II systematic deviation from the nominal concentration P

constant segments ΔL, then this process corresponds to a stratified sample of proportional sub-division and hence, if the length ΔL is small compared to the total lenght L, the expected total variance is given by:

$$\sigma^2_{tot} = E(\sigma^2_z(r,t)) + \sigma^2_{syst}(P(r,t)) \qquad (12.1.1)$$

The total variance is composed of the expectation of the variances within the strata, $E(\sigma^2_c(r,t))$, and the variance of expectations, $\sigma^2_{syst}(P(r,t))$. The index c is meant to express that the variances within the "strata" L are of the chance (random) type only, while the variance σ^2_{syst} describes the mean systematic concentration distribution P(r,t).

$\sigma^2_c(r,t)$ is given by the variance of the random mixture which has a nominal concentration P(r,t) at the position r at time t:

$$\sigma^2_z(r,t) = P(r,t)(1 - P(r,t)) \cdot \frac{g_1}{g} \qquad (12.1.2)$$

Müller[31] was the first to show, for constant density under the condition $P(r_j,t) \ll 1$, that the spatial expectation of $\sigma^2_c(r,t)$ can be approximated by:

$$E(\sigma^2_z(r,t)) = \sigma^2_z - \frac{g_1}{g}\sigma^2_{syst}(t) \qquad (12.1.3)$$

σ^2_c is here the time-independent variance of the random mixture having the nominal value P. It follows from equation (12.1.2) that (12.1.3) holds exactly and without limitation. The following relationship is hence deduced for the variance σ^2_{tot} of a non-random mixture:

$$\boxed{\sigma^2_{tot} = P(1-P) \cdot \frac{g_1}{g} + \left(1 - \frac{g_1}{g}\right) \cdot \sigma^2_{syst}} \qquad (12.1.4)$$

In the case of total segregation, $\sigma^2_{syst} = P(1-P)$ and hence:

$$\sigma^2_{tot} = P \cdot (1-P) = \sigma^2_{syst} \qquad (12.1.5)$$

ie. in the case of complete segregation all random fluctuations disappear and the total variance agrees with the systematic variance. In the other extreme case of pure random mixture, every systematic concentration deviation disappears: $\sigma^2_{syst} = 0$. The total variance is identical to the variance of the random mixture:

$$\sigma^2_{tot} = P(1-P) \cdot \frac{g_1}{g} = \sigma^2_z \qquad (12.1.6)$$

In order to obtain a confidence probability for the sample concentration X as an estimate for the nominal concentration P, the variance σ^2_{syst} and hence also the variance σ^2_{tot} must be known or be experimentally determined. The variance which is given in the case of complete segregation is the largest possible. Although it can be used as an upper estimate in the case of a completely unknown state of mixing, it usually produces a disproportionately large confidence region, in particular if the sample size g is large compared to the particle size g_1. In the extreme case in which the sample size agrees with the particle size the total variance nevertheless corresponds to the variance of a mixture in which the components are completely separate.

12.2 Correlation length, correlogram

In the case of non-random mixtures whole regions of the material being mixed deviate systematically from the expected overall composition of the material. If samples are taken from the mixture the concentrations x_i, x_j of "closely adjacent" samples are not independent of one another. The degree of dependence of two variables is described by the covariance or by the correlation coefficient ρ. The correlation coefficient depends on the distance between the samples to be compared. If the distance is very great, then the dependence disappears and the covariance becomes zero. In this case, the variance of the difference of the two sample concentrations is given by:

$$\sigma^2(x_i - x_j) = \sigma^2(x_i) + \sigma^2(x_j) = 2\,\sigma^2(x) \tag{12.2.1}$$

However, if the variables are correlated we have in general:

$$\sigma^2(x_i - x_j) = \sigma^2(x_i) + \sigma^2(x_j) - 2\,\mathrm{Cov}\,(x_i\,x_j) \tag{12.2.2}$$

or expressed with the correlation coefficients ρ_{ij} between the sample concentrations x_i and x_j:

$$\sigma^2(x_j - x_j) = 2\,\sigma^2(x)\,(1 - \rho_{ij}) \tag{12.2.3}$$

If it is assumed that ρ_{ij} depends only on the distance between two samples, then those can be selected from the total number of possible samples which are an identical distance apart and estimates r_{ij} can hence be calculated. Let n be the total number of all samples which are drawn in the sequence x_1, x_2 ....x_n and k <<n the "distance" of the (n-k) pairs having the concentrations (x_r, x_{r+k}), where r = 1,2, ... (n-k). The sample covariance is thus given by

$$\mathrm{Cov}\,(x_v, x_{v+k}) = \frac{1}{n-k} \sum_{r=1}^{n-k} (x_r - P)(x_{r+k} - P) \tag{12.2.4}$$

Let the following hold for the sample variance:

$$s_x^{\,2} = s^2(x_r) = s^2(x_{r+k}) = \frac{1}{n} \sum_{l=1}^{n} (x_l - P)^2$$

The estimate for the correlation coefficients r_k is therefore given by:

$$r_k = \frac{\frac{1}{n-k} \sum_{r=1}^{n-k} (x_r - P)(x_{r+k} - P)}{\frac{1}{n} \sum_{l=1}^{n} (x_l - P)^2} \qquad (12.2.5)$$

If the sample variance s_d^2 for the differences $d = x_{r+k} - x_k$ of the measured values x_r, x_{r+k} at a distance k is given by:

$$s_d^2 = \frac{1}{n-k} \sum_{r=1}^{n-k} (x_r - x_{r+k})^2 \qquad (12.2.6)$$

then correspondingly to equation (12.2.3) we also have:

$$r_k = 1 - \frac{1}{2} \left(\frac{s_d^2}{s_x^2} \right) \qquad (12.2.7)$$

Figure (12.2.1) shows the curve of an observed correlation r_k plotted against the spatial distance l_k between the samples[32]. By definition $r_{k=0} = 1$ for the distance $l_k = 0$. Adjacent samples are still highly correlated, for example for $l_k = 1"$, r_k 0.5. The correlation decreases rapidly with increasing distance. For distances greater than $l_n = 5"$, the values for r_k are no longer significantly different from 0. l_n is termed the correlation

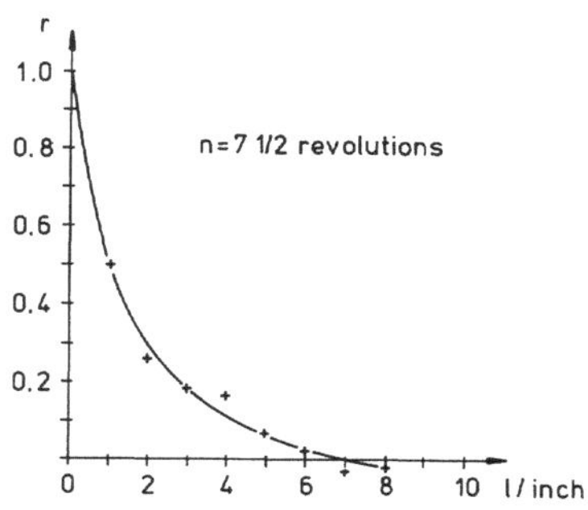

Figure 12.2.1 Correlogram in a Y mixer

length λ. The correlation length is a measure of the degree of dependence of adjacent samples and hence a measure for mixtures deviating from the pure random state. Lacey has shown that the correlation length in a Y mixer can also depend on which direction is being considered. These investigations hence permit inferences about the nature of mixing mechanismus.

13 Sampling from a Convevor Belt

13.1 Variance within and between belt sections of lenght L

Stange[33] was the first to point out that the systematic distribution of characteristics must also be taken into account in sampling from a conveyor belt. It is presumed on the basis of what is known about the preparation process that ash contents or metal contents of "closely adjacent" samples are not independent of one another. Values of samples are therefore correlated to a degree which depends on the distance between them. The characteristics X_1, X_2 ... X_n of elements lying on a belt fluctuate about the expectation P with a total variance of σ^2_{tot}. The correlation coefficient $\rho(y)$ existing between two elements a distance yapart forms a straight line function of the distance y (Figure 13.1.1). The correlation extends in both directions symmetrically over a finite range characterized by the correlation length λ. Outside this range $\rho(y) = 0$ for $|y| \geq \lambda$. Let a belt section of length L contain N elements lying on it. The elements are numbered in sequence.

Concentrations X_r of the individual elements can also be considered to be of the form

$$X_r = P + \varepsilon_r \tag{13.1.1}$$

in which ε_r represents a deviation from the total composition expected. These deviations have an expectation:

$$E(\varepsilon_r) = 0 \tag{13.1.2}$$

and a variance:

$$\sigma^2(\varepsilon_r) = \sigma^2(X_r - P) = \sigma^2(X_r) = \sigma^2_{tot} \tag{13.1.3}$$

Table 13.1.1 Random sampling from a belt section of length L

Task: What is the size of the sampling error when taking n samples randomly from a belt section of length L?

1st step Determination of the total variances σ_{tot}^2 which would be expected on taking all elements from the whole belt. σ_{tot}^2 may be estimated by prior experiments with samples over all belt sections, or it may be known by long experience or may be given by the nature of the material (binomial distribution, Poisson distribution).

2nd step Determination of the correlation function $\rho(y)$ by k prior experiments for each sample distance y:

$$\rho(y) = \frac{\text{Cov}(X_a; X_{a+y})}{\sigma_{tot}^2} \approx \frac{\frac{1}{k}\sum_{i=1}^{k} X_{a_i} \cdot X_{a_i+y} - \frac{1}{k}\sum_{i=1}^{k} X_{a_i} \cdot - \frac{1}{k}\sum_{j=1}^{k} X_{a_j+y}}{\sigma_{tot}^2}$$

3rd step Determination of the correlation length $y = \lambda$ for $\rho(y > \lambda) = 0$

4th step Calculation of the mean correlation coefficient ρ_m

(13.1.15) $$\rho_m = \frac{1}{\lambda} \int_0^{\lambda} \rho(y)\, dy$$

5th step Selection of a belt length section L, in which $L \gg \lambda$.

6th step Random taking of n samples from a belt length section L and calculation of the mean $\bar{X} = \Sigma X_i/n$

$$\bar{X} = \frac{1}{n}\sum_{i=1}^{n} X_i$$

7th step Calculation of the sampling error of $\bar{X}$:

(13.2.8) $$\boxed{\sigma_{tot}^2(\bar{X}) = \frac{\sigma_{ges}^2}{n} + 2\,\sigma_{tot}^2 \cdot \frac{\lambda}{L} \cdot \rho_m}$$

Table 13.1.2 Systematic sampling from a belt section of length L

Task: What is the size of the sampling error when taking n samples systematically from a belt section of length L?

1st step Determination of the total variance σ_{tot}^2 which would be expected on taking all elements from the whole belt. σ_{tot}^2 may be estimated by prior experiments with samples over all belt sections, or it may be known by long experience or may be given by the nature of the material (binomial distribution, Poisson distribution).

2nd step Choice of a belt length section L and of the number of samples to be investigated n << N (N is the total number of possible samples)

3rd step Taking of n samples, at a distance $l = L/n$ apart, from the belt section chosen and calculation of the mean $\bar{X} = \Sigma X_i/n$

4th step Determination of correlation functions with the distance $i \cdot l$, $i = 1, 2, \ldots n$ from prior experiments:

$$\rho_{i\cdot l} = \rho(i\cdot l) = \frac{\mathrm{Cov}(X_a;\, X_{a+i\cdot l})}{\sigma_{tot}^2} \approx \frac{\frac{1}{n}\sum_{\nu=1}^{n} X_{a_\nu} \cdot X_{a_\nu+i\cdot l} - \frac{1}{n}\sum_{\nu=1}^{n} X_{a_\nu} \cdot \frac{1}{n}\sum_{\mu=1}^{n} X_{a_\mu+i\cdot l}}{\sigma_{tot}^2}$$

5th step Calculation of the sampling error for

$$\sigma^2(\bar{X}) = \frac{\sigma_{tot}^2}{n} + \sigma_{tot}^2 \left(\frac{2l}{L} \sum_{i=1}^{n} \left(1 - \frac{i\cdot l}{L}\right) \cdot \rho_{i\cdot l}\right)$$

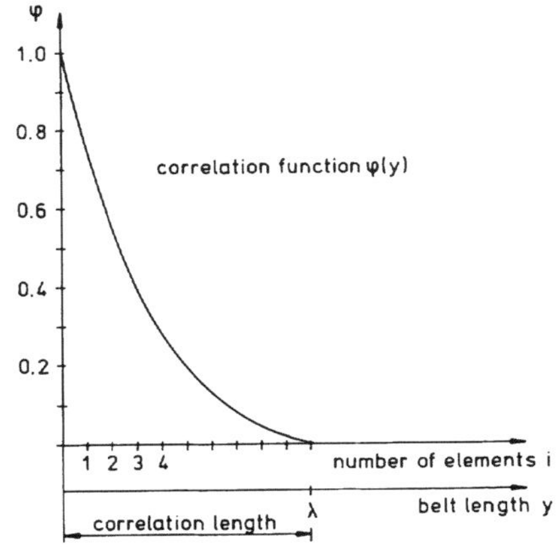

Figure 13.1.1 Schematic illustration showing the definition of correlation length and of correlation function

Since the correlation depends only on the distance of the sample considered:

$$\text{Cov}\ (\varepsilon_r \cdot \varepsilon_{r+k}) = \sigma^2_{tot} \cdot \rho_k \tag{13.1.4}$$

a mean of all values can be obtaind for the N elements in the belt length section L:

$$P_L = \frac{1}{N}\sum_{i=1}^{N} X_i \tag{13.1.5}$$

Since the N elements only constitute a sample from the whole belt, the mean P_L can be different from the expectation P.

The variance σ^2_L of element characteristics X_i within the belt length section is given by:

$$\sigma^2_L = \frac{1}{N} \sum_{i=1}^{N} (X_i - P_L)^2 \tag{13.1.6}$$

The belt length section L can be considered as a "stratum" of the total belt in the sense of a stratified sample. If, as is customary in industrial sampling from a conveyor belt, samples are systematically taken at specific time intervals, and hence, for a constant conveyor belt speed, at constant distances, the sample size is proportional to the belt length section L. On proportional subdivision of a stratified sample we have:

$$\sigma_{tot}^2 = E(\sigma_L^2) + \sigma^2 (P_L) \tag{13.1.7}$$

The total variance is composed of the expectation of the variances σ_L^2 within the belt length sections and the variance of the expectations P_L between the belt length sections.

Variance between belt sections

$$\sigma^2(P_L) = \sigma^2(\frac{1}{N}\sum_{i=1}^{N} X_i) = \frac{1}{N^2}\ \sigma^2(\sum_{i=1}^{N} X_i)$$

$$\sigma^2(P_L) = \frac{1}{N^2} \cdot \sigma^2(\sum_{i=1}^{N} \varepsilon_i) \tag{13.1.8}$$

The ε_i are not independent. Hence:

$$\sigma^2(P_L) = \frac{1}{N^2} \left\{ \sum_{i=1}^{N} \sigma^2(\varepsilon_i) + \sum_{i \neq j}^{N(N-1)} \mathrm{Cov}\,(\varepsilon_i, \varepsilon_j) \right\}$$

$\varepsilon^2(\varepsilon_i)$ is identical to σ_{tot}^2, according to equation (13.1.3). Elements at constant distances from one another can always be collected together for the purpose of calculating the covariances. There are (N-1) immediately adjacent elements, (N-2) elements which have one element in between, etc. If it is also taken into account that $\mathrm{Cov}(\varepsilon_i, \varepsilon_j)$) $\mathrm{Cov}(\varepsilon_j, \varepsilon_i)$, then with the aid of equation

(13.1.4) $\sigma^2(P_L)$ is calculated to be:

$$\sigma^2(P_L) = \frac{\sigma^2_{tot}}{N} + \frac{2\,\sigma^2_{tot}}{N^2}\,(\,(N-1)\,\rho_1 + (N-2)\cdot\rho_2 + \ldots\,)\,)$$

or

$$\sigma^2(P_L) = \frac{\sigma^2_{tot}}{N} + \frac{2\,\sigma^2_{tot}}{N}\sum_{i=1}^{N}(1 - \frac{i}{N})\cdot\rho_i \qquad (13.1.9)$$

$\sigma^2(P_L)$ is the variance of the mean determined from all elements of a belt section.

For practical application, the summation in equation (13.1.9) is replaced by an integral in which the expression i/N is approximated by a standardized positional coordinate y/L.

$$\frac{i}{N} = \frac{y}{L} \qquad (13.1.10)$$

Equation (13.1.9) can therefore be rewritten as:

$$\sigma^2(P_L) = \frac{\sigma^2_{tot}}{N} + \frac{2\,\sigma^2_{tot}}{L}\int_{\varepsilon\to 0}^{L}(1 - \frac{y}{L})\cdot\rho(y)\,dy \qquad (13.1.11)$$

y is by definition a distance and can therefore only assume positive values. The lower integration limit ε corresponds exactly to one "element distance". When these element sections are made arbitrarily small during the transition from the summation to the integral, ε can tend arbitrarily close to zero, but theoretically it can never quite get there. Since this case is of importance only for completely uncorrelated samples, the lower integration limit is always set equal to zero in any further considerations. The case of uncorrelated samples is treated separately.

Special cases:

1) If there are no systematic but only random deviations on the belt, $\varepsilon(y) = 0$ for all $y \neq 0$. The variance $\sigma^2(P_L) = \sigma^2_{tot}/N$ is thus identical to the variance of a random sample having N elements.

2) If the belt length L is large compared with the correlation length λ, then the area of integration above the correlation length is adequate for calculating the integral. The following inequality holds:

$$1 \geq 1 - \frac{y}{L} \geq 1 - \frac{\lambda}{L} \qquad (13.1.12)$$

Da $\lambda/L \ll 1$ and since $\lambda/L \ll 1$ is assumed we have:

$$1 - \frac{y}{L} \approx 1 \qquad (13.1.13)$$

The correlation integral is thus simplified and leads to:

$$\boxed{\sigma^2(P_L) = \frac{\sigma^2_{ges}}{N} + \frac{2\lambda}{L} \cdot \rho_m \cdot \sigma^2_{tot}} \qquad (13.1.14)$$

where:

$$\rho_m = \frac{1}{\lambda} \cdot \int_0^L \rho(y)\, dy \qquad (13.1.15)$$

In the case of a systematic concentration distribution having a positive correlation the variance $\sigma^2(P_L)$ is always greater than the variance of a random mixture for a sample having the same number N of elements. In order to keep the variance as small as possible the belt length section to be investigated has to be considerably greater than the correlation length.

Example:

Stange[9] gives as an example these values obtained in a series of experiments on a conveyor belt in a coal cleaning plant. The sample size required for a determination of the ash content is 3 lb. The correlation length λ corresponds to an amount of 400 t. The mean correlation coefficient was found to be $\rho_m = 0.14$.

For an investigated belt lenght of $L = 7\lambda$ whis corresponds to a number of almost $N \cong 10^6$ samples of 3 lb each. Investigation of all of these samples without correlation (random mixture) would produce a value of:

$$\sigma^2(P_L) = 10^{-6} \cdot \sigma^2_{tot}$$

However, if the correlation $\rho_m = 0.14$ is taken into account:

$$\sigma^2(P_L) = 10^{-6} \cdot \sigma^2_{tot} + \frac{2}{7} \cdot 0{,}14 \cdot \sigma^2_{tot}$$

$$\sigma^2(P_L) = 4 \cdot 10^{-2} \cdot \sigma^2_{tot}$$

In view of the huge number of samples investigated this value is very large. The comparison shows that in the presence of a correlation the assumption of a random mixture can produce false estimates which are far too good.

Variance within belt sections

Equations (13.1.7) and (13.1.11) permit the calculation of the variance expected within belt length sections of size L. The assumption $1/N \ll 1$ leads to:

$$E(\sigma^2_L) = \sigma^2_{tot} \left\{ 1 - \frac{2}{L} \int_0^L \left(1 - \frac{y}{L}\right) \cdot \rho(y)\, dy \right\} \qquad (13.1.16)$$

The expectation for variances within the strata is, for an existing correlation, always less than the total variance.

Special cases

1) If there are no systematic deviations, then $E(\sigma_L^2)$ is identical to the variance of the population.

2) If the belt length L is large compared to the correlation length λ, then the equation can again be simplified by assuming $(1 - y/L) \cong 1$:

$$E(\sigma_L^2) = \sigma_{tot}^2 \left\{ 1 - \frac{2\lambda}{L} : \rho_m \right\} \quad (13.1.17)$$

The meaning of ρ_m has already been made clear by equation (13.1.15).

For a sufficiently large belt length the quantity $2(\lambda/L).\rho_m$ can be neglected compared with 1 and we have:

$$E(\sigma_L^2) = \sigma_{tot}^2$$

In this case the variance within a stratum is a good estimate of the total variance.

Example

In the example given above, the expectation $E(\sigma_L^2)$ was estimated by $\sigma_L^2 = 0.28 \times 10^{-4}$ ash content for a belt section of $L \cong 23\ \lambda$. For $\rho_m = 0.14$, $2(\lambda/L) . \rho_m \cong 0.01$ and negligible compared with 1 . σ_L^2 is hence also a good estimate of the total variance.

13.2 Inspection of given quantities when sampling randomly or systematically from a conveyor belt

Up to now, all elements of a belt length section L have been included in the investigation. In practice this is usually unrealistic as may be seen from the large number of elements (10^6 samples in the above example). It is precisely the aim of sampling to reduce this effort by investigating only a few samples. In order to estimate the composition of a prescribed belt length, n individual samples having properties X_1, X_2 ... X_n are taken from the flow of material. The estimate required is found by taking the mean of these n values.

$$\bar{X} = \frac{1}{n}\sum X_i \qquad (13.2.1)$$

All n samples are elements of a belt length section of length L. It is quite possible to take these samples by various methods. For one, by a purely random taking of samples, for example with the aid of random number tables, and for another by a systematic taking of samples, preferred in practice, at constant distances l, only the position of the first element being determined randomly. In both cases, the population of all N elements in the belt length section can be considered as a population of its own having an expectation P_L (equation 13.1.5) and a variance σ_L^2 (equation 13.1.6).

Random sample

In a random sample a total sample of size n is removed from this population. For a random sample without replacement this means that:

$$\sigma_L^2(\bar{X}) = \frac{N-n}{N-1} \cdot \frac{\sigma_L^2}{n} \approx (1 - \frac{n}{N}) \cdot \frac{\sigma_L^2}{n} \qquad (13.2.2)$$

$$E_L(\bar{X}) = P_L \qquad (13.2.3)$$

The variance $\sigma_L^2(\bar{X})$ fluctuates from belt length section to belt length section. The expectation is given by:

$$E(\sigma_L^2(\bar{X})) = (1 - \frac{n}{N}) \cdot \frac{E(\sigma_L^2)}{n} \qquad (13.2.4)$$

The expectation $E(\sigma_L^2)$ was calculated in section (1) by means of equation (13.1.16). Hence:

$$\boxed{E(\sigma_L^2(\bar{X})) = (1 - \frac{n}{N}) \cdot \frac{\sigma_{tot}^2}{n} (1 - \frac{2}{L} \int_0^L (1 - \frac{y}{L}) \cdot \rho(y)\, dy)} \qquad (13.2.5)$$

The expectation $E(\sigma_L^2(\bar{X}))$ is for a given positive correlation always smaller than the corresponding value in the absence of correlation. In an extreme case characteristic values within belt length sections are all equally large but vary from one belt section to another. The correlation integral is in this case exactly 1 and $E(\sigma_L^2(\bar{X})) = 0$. Although this means that unambiguous statements can be made about the values P_L on the basis of means $\bar{X}$, they cannot be made about the expectation P of the population of the belt, which expectation is what we are interested in. In going on to consider the population of the total belt having the variance σ_{tot}^2, the values of equations (13.2.2) and (13.2.3) represent conditional means for variances. The variance of $\bar{X}$ is calculated for this superordinate population by means of equation (2.9.17) as:

$$\sigma_{tot}^2(\bar{X}) = E(\sigma_L^2(\bar{X})) + \sigma^2(E_L(\bar{X})) \qquad (13.2.6)$$

By means of equations (13.2.5) and (13.1.16) it follows that:

$$\sigma_{tot}^2(\bar{X}) = (1 - \frac{n}{N}) \cdot \frac{\sigma_{tot}^2}{n} (1 - \frac{2}{L} \int_0^L (1 - \frac{y}{L}) \cdot \rho(y)\, dy)$$

$$+ \frac{\sigma_{tot}^2}{N} + \frac{2\, \sigma_{tot}^2}{L} \int_0^L (1 - \frac{y}{L}) \cdot \rho(y)\, dy$$

If n and N are large compared with one, then $\sigma^2_{tot}(\bar{X})$ is given by:

$$\sigma^2_{tot}(\bar{X}) \approx \frac{\sigma^2_{tot}}{n} + \frac{2\,\sigma^2_{tot}}{L} \int_0^L (1 - \frac{y}{L}) \cdot \rho(y)\, dy \qquad (13.2.7)$$

The variance of X in the population is for a given positive correlation always larger than for a purely random distribution of characteristic values. If the sample size n is very large, then $\sigma^2_{tot}(X)$ tends to a constant value.

Special cases:

1) If elements are independent of one another, ie. the correlation function $\rho(y) = 0$, then the totál variance corresponds to that of a random sample, $\sigma^2_{tot}(\bar{X}) = \sigma^2_{tot}/n$.

2) If the concentrations within a belt length section are all equally large, and if it is also assumed that the sample size n_L is very large, then the correlation integral $(2/L)\int(1-y/L)\rho(y)dy$ has the value one and it follows for the total variance that: $\sigma^2_{tot}(\bar{X}) = \sigma^2_{tot}$. This result is readily understood. Repeated sampling within a belt section does not produce any new information and the mean $\bar{X}$ therefore also produces no higher precision.

3) If the correlation length $\lambda \ll L$, then the mean correlation coefficient ρ_m in accordance with equation (13.1.15) can be introduced and we have:

$$\sigma^2_{tot}(\bar{X}) = \frac{\sigma^2_{tot}}{n} + 2\,\sigma^2_{tot} \cdot \frac{\lambda}{L} \cdot \rho_m \qquad (13.2.8)$$

Systematic sampling

In systematic sampling, the n elements of a sample are taken as already mentioned at a constant distance $l = L/n$. If the belt length section of length l contains exactly $k = N/n$ elements, then there are k different ways of starting the sampling. Each possibility can lead to different means $\bar{X} = (1/n) \sum X_i$. If the belt section of length L is considered as a population on its own having an expectation P_L and a variance σ_L^2, then:

$$\sigma_L^2(\bar{X}) = \sigma^2\left(\frac{1}{n}\sum_{i=1}^{n} X_i\right) = \frac{\sigma_L^2}{n} + \frac{1}{n^2}\sum_{i \neq j}^{n(n-1)} \mathrm{Cov}(X_i, X_j)$$

Since the samples are being removed systematically, the elements are no longer independent and covariances have to be taken into account. The distance between any element and the next one following it is l. There are thus (n-1) elements at a distance l, (n-2) elements at a distance 2 l etc. If it is taken into account that $\mathrm{Cov}(X_i, X_j) = \mathrm{Cov}(X_j, X_i)$ and if further the correlation coefficient depends only on the distance between particular elements, then:

$$\sigma_L^2(\bar{X}) = \frac{\sigma_L^2}{n} + \frac{2\,\sigma_{tot}^2}{n}\sum_{i=1}^{n}\left(1 - \frac{i \cdot l}{L}\right) \cdot \rho_{i \cdot l} \qquad (13.2.9)$$

Here the summation cannot in general be replaced by an integral since the elements removed are not close to one another. The variance $\sigma_L^2(X)$ varies from belt length section to belt length section. The expectation is given by:

$$E(\sigma_L^2(\bar{X})) = \frac{E(\sigma_L^2)}{n} + \frac{2\,\sigma_{tot}^2}{n}\sum_{i=1}^{n}\left(1 - \frac{i \cdot l}{L}\right) \cdot \rho_{i \cdot l} \qquad (13.2.10)$$

With the aid of equation (13.1.11) obtain further:

$$E(\sigma_L^2(\bar{X})) = \frac{\sigma_{tot}^2}{n}\left(1 - \frac{2}{L}\int_0^L \left(1 - \frac{y}{L}\right)\rho(y)\,dy\right) + \frac{\sigma_{tot}^2}{n} \cdot 2\sum_{i=1}^{n}\left(1 - \frac{i \cdot l}{L}\right) \cdot \rho_{i \cdot l}$$

or with n = L/1

$$E(\sigma_L^2(\bar{X})) = \frac{\sigma_{tot}^2}{n} + \sigma_{tot}^2 \left(\frac{2\,1}{L} \sum_{i=1}^{n} \left(1 - \frac{i \cdot 1}{L}\right) \cdot \rho_{i \cdot 1} - \frac{2}{L} \int_0^L \left(1 - \frac{y}{L}\right) \rho(y)\, dy\right) \qquad (13.2.11)$$

In the extreme case all concentrations within a belt section are of the same size. In this case, the correlatin integral $2/L \int (1-y/L) \cdot \rho(y)\, dy$ is again exactly equal to 1.

The summation is expanded as follows:

$$\frac{2}{n} \sum_{i=1}^{n} \left(1 - \frac{i}{n}\right) = \frac{2}{n}\left(n - \frac{n+1}{2}\right) = \frac{n-1}{n} \qquad (13.2.12)$$

This means that the expectation of $\sigma_L^2(\bar{X})$ is given by:

$$E(\sigma_L^2(\bar{X})) = \frac{\sigma_{tot}^2}{n} + \sigma_{tot}^2 \left(\frac{n-1}{n} - 1\right) = 0 \qquad (13.2.13)$$

This result has already been obtained for a pure random selection from within a belt length section. If the concentrations within a belt section are all of the same size it does not matter whether the sample is taken systematically or randomly.

Although statements about the values P_L are possible by means of $\sigma_L^2(\bar{X})$, they cannot be made about the expectation P required of the total belt. Analogously to the case of random selection $\sigma_L^2(\bar{X})$ represents a conditional variance. The variance of $\bar{X}$ in the superordinate population is given by:

$$\sigma_{tot}^2(\bar{X}) = E(\sigma_L^2(\bar{X})) + \sigma^2(E_L(\bar{X}))$$

By means of equations (13.2.11) and (13.1.11) it follows that:

$$\sigma^2_{tot}(\bar{X}) = \frac{\sigma^2_{tot}}{n} + \sigma^2_{tot}\left(\frac{2l}{L}\sum_{i=1}^{n}\left(1 - \frac{i\cdot l}{L}\right)\cdot \rho_{i\cdot l} - \frac{2}{L}\cdot\int_0^L \left(1 - \frac{y}{L}\right)\cdot \rho(y)\,dy\right)$$

$$+ \frac{\sigma^2_{tot}}{N} + \frac{2\,\sigma^2_{tot}}{L}\int_0^L \left(1 - \frac{y}{L}\right)\cdot \rho(y)\,dy$$

or with $N \gg n$

$$\boxed{\sigma^2_{tot}(\bar{X}) = \frac{\sigma^2_{tot}}{n} + \sigma^2_{tot}\left(\frac{2l}{L}\sum_{i=1}^{n}\left(1 - \frac{i\cdot l}{L}\right)\cdot \rho_{i\cdot l}\right)} \qquad (13.2.14)$$

The variance $\sigma^2_{tot}(\bar{x})$ for systematic sampling has an appearance similar to that of the variance for random sampling from within a belt length section (cf. equation (13.2.7)). The correlation sum replaces the correlation integral. However, as n increases the summation tends toward the integral and the variance $\sigma_{tot}^2(\bar{x})$ towards the same constant value.

<u>Special cases</u>

1) If the elements are independent of one another, systematic sampling agrees with random sampling:
$\sigma^2_{tot}(\bar{x}) = \sigma^2_{tot}/n$

2) If the distance l is larger than the correlation length λ, then sampling is identical to that of random sampling, since $\rho_{i.l}$ for $l \gg \lambda$. Since in the case of a positive correlation this produces at the same time the smallest variance for a given n, this state should always be aimed at.

3) If the concentrations within a belt length section are all of the same size, then the correlation sum $(2l/L)\ \Sigma\ ((1-i.l/L)\ \rho_{i.l})$ has the value $(n-1)/n$ according to equation (13.2.12) and the

variance is given as $\sigma^2_{tot}(\bar{X}) = \sigma^2_{tot}$. Just as in the case of random sampling, the large number n of elements in the sample does not produce any new information in this case. The precision cannot therefore be improved by increasing the number of elements. This state of affairs tends to manifest itself even for correlation coefficients not equal to one. In practice, this means that in the case of a high correlation the sample size can be reduced without significant loss of precision of the statement.

14 Sampling Devices

Sampling, splitting and analysis used to be carried out only manually. Increasing automation and rationalization lead more and more to integrated automatic sampling systems, which are frequently specifically adapted to the materials to be investigated and to the degree of precision required. This trend has produced and is still producing a multitude of sampling devices. It is therefore impossible, within the scope of this paper, to aim at providing comprehensive coverage of these devices. Instead, a few sampling devices and sample-splitting devices for various material systems, such as liquids, gases, suspensions or aggregates of solid particles, are introduced as examples. For the same reason, no attempt is made to list the sample-preparation devices which are frequently necessary, such as comminution machine, mixer, dryer and the like.

14.1 Devices for sampling liquids

In the laboratory, a pipette is still the most frequently used device for sampling stationary liquids. On an industrial scale, the so-called open plunging syphon (Figure 14.1.1.a), also called a sampler, corresponds to the pipette. An open plunging syphon essentially comprises a tube of about 20 - 30 mm internal diameter, which is constricted near the top and the bottom. The open plunging syphon is immersed to the depth desired. After the upper opening has been closed, the sample can be removed from the container. A sealable plunging syphon (Figure 14.1.1.b) consists of a tube with a handle which is connected to the lower sealing disc via a tension spring. The tube can be completely sealed after the liquid has flowed in. The sealable plunging syphon B (Figure 14.1.1.c) also consists of a cylinder 4 with a protecting ring 3 and a tube extension 6 which is used if required. The rod 2 and, with it, the sealing pistons 5 are pulled upwards with the aid of a pull-ring 8 and the chain 7. The liquid flows in until the conical sealing stopper 1 closes the tube to give a tight seal[34].

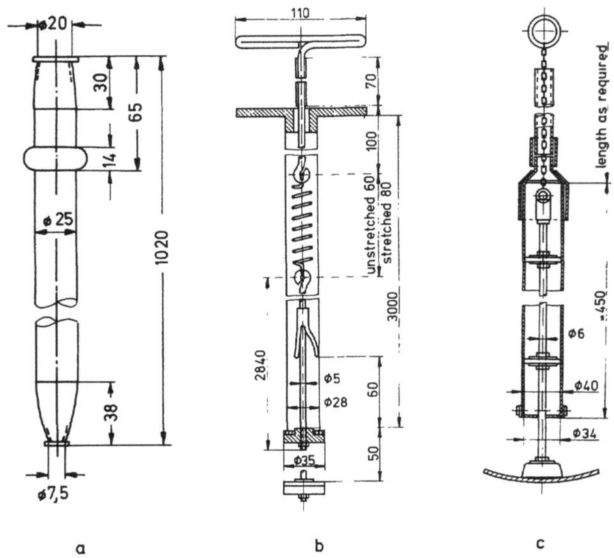

Figure 14.1.1 Plunging syphons according to DIN 51.750, Sheet 2

Open plunging syphons are mainly suitable for taking samples of liquids from small containers and barrels. Sealable plunging syphons are used especially for taking layer samples from inhomogeneous stationary liquids.

In contrast to the plunging syphon, the immersion flask (Figure 14.1.2) or immersion bomb enables layer samples to be taken from even comparatively large containers, such as tank cars, ships' tanks, or tanks on land. The flask is a lead-weighted metal bottle, which is sealed with a plug and lowered into the liquid. A valve cord enables the valve to be opened with a short, sharp pull.

When sampling natural waters, it is frequently desired to take average samples over a relatively long period, as can be done, for example, by means of a Passavant DBGM immersion collector[35] (Figure 14.1.3). A 2-liter sample is taken in a continuous manner over a 24-hour period at an immersion depth of 0.5 m.

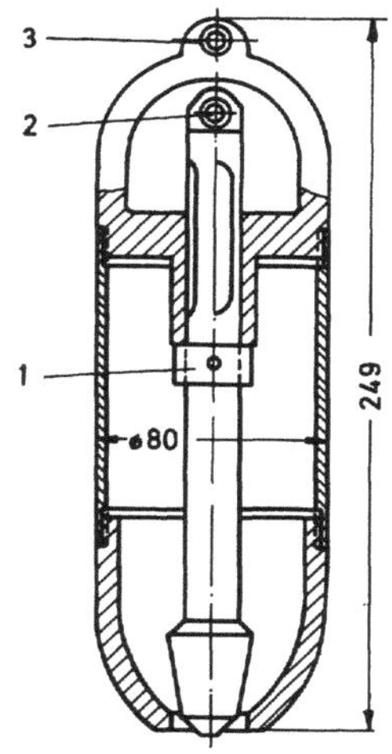

Figure 14.1.2 Immersion flask according to DIN 51.750, Sheet 2 1 Valve spindle, 2 Attachment for valve pull-cord, 3 Attachment for the chain on which the immersion flask is lowered

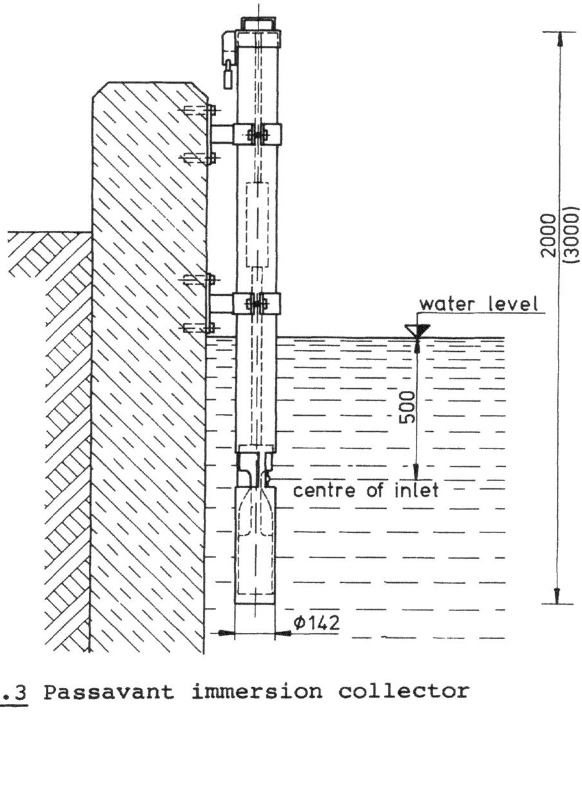

Figure 14.1.3 Passavant immersion collector

The sample is forced, by the static pressure of the liquid above the sampling point, into a glass bottle, the air in the bottle gradually escaping through a capillary. The device is enclosed by a steel tube in order to protect it from damage.

Sampling from a stationary suspension is always particularly difficult when the particles settle and cannot be kept in suspension even by constant stirring. Inflowing can also cause local demixing at the inflow opening, which can render the result misleading. Fewer difficulties are experienced in the case of flowing suspensions. If possible, sampling is carried out at the outflow opening, for example by means of scoops, even in the case of liquid metals or slags, or by means of slotted vessels which have an opening in the shape of a rectangular slot. The vessel is drawn through the outflowing liquid (or suspension). In the case of slurries with low solids contents, which do not tend to segregate, the sample can be taken from a part stream (outlet cock) or via a sampling valve. Figure 14.1.4 shows two Strahman sampling valves[36)], one open and the other closed.

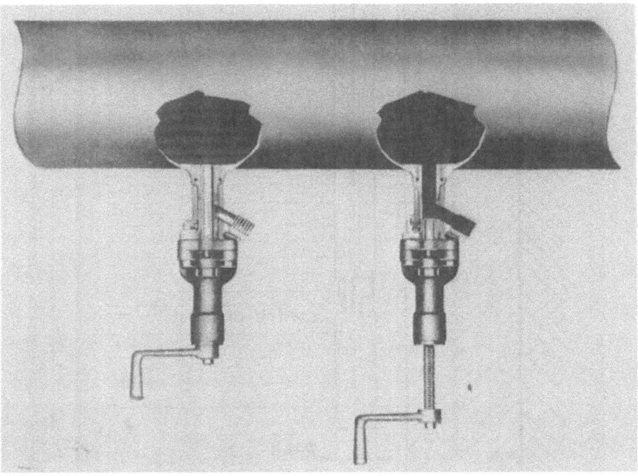

<u>Figure 14.1.4</u> Strahman sampling valve

14.2 Devices for sampling gases and dusts

Among the devices for determining the gas contents in air, some are suitable for carrying out analyses direct on the spot. In the multi-gas detector marketed by Messrs. Drägerwerk[37], a precisely fixed amount of impure air is pumped through a so-called Dräger tube by means of a simple bellows pump (Figure 14.2.1). The formulations contained in the tube react by becoming discolored and hence allow inferences to be drawn regarding the composition of the air sample.

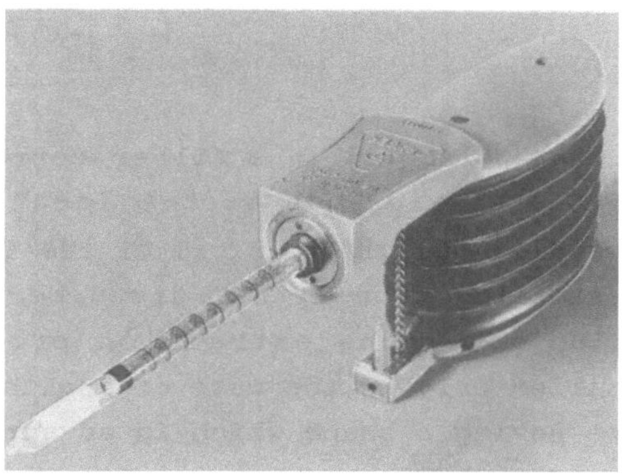

Figure 14.2.1 Multi-gas detector

If the analysis can only be carried out in a laboratory, air samples can be preserved in vessels, so-called gas sample tubes, resembling the immerison flask previously mentioned. The vessles draw in air by means of a vacuum pump, after which they are sealed and forwarded to a laboratory for investigation.

In the case of flowing gases, a part stream is drawn off. If dust particles dispersed in the gas are to be investigated, they have to be separated from the part stream by means of filters or dust separators. Figure 14.2.2 shows the layout of a conventional arrangement for sampling dust-laden gases[38].

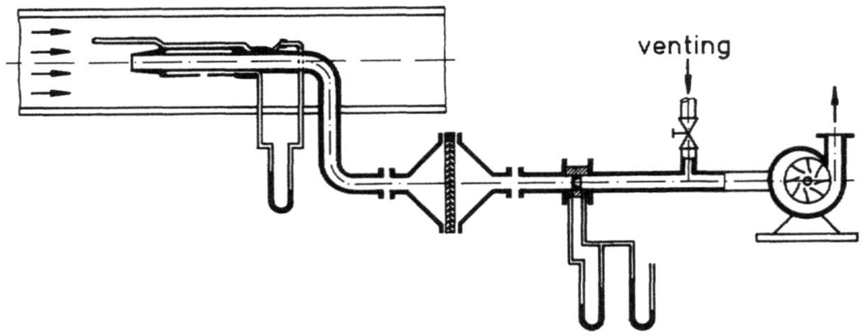

Figure 14.2.2 Arrangement for sampling inflowing dust-laden gases 1 Sampling tube, a and b manometric probes, 2 filter, 3 orifice plate, 4 fan

The quantity of dust collected in a filter corresponds to the dust content of the carrier gas at the location where the measurement is carried out, if the form of the flow streamlines upstream of the sampling tube is not disturbed either by the probe installation or by the extraction process[39]. This condition is fulfilled only in the case of isokinetic extraction by means of probe having a shape which is advantageous in terms of fluid mechanics. Rüping[40] reports on experiments to determine the influence of extraction errors and their effects on the measurements of the amounts of dust in the flowing gases.

As already mentioned in the case of sampling from liquids, local demixing caused by flow pattern disturbances is far more frequent still in the case of gases. Sampling conditions are ideal when the flow is completely undisturbed. The optical analysis of the particles by means of scattered light is one such ideal method. It is desirable to be able to use this method for measurements even at very high particle concentrations. The scattered light measurement should be carried out on individual particles in order to avoid having to make assumptions about the nature of the particle size distrubution of the dust.
Umhauer[41] has presented a device satisfying these aspects, by means of which measurements are possible up to concentrations

of 10^6 particles/cm^3. The lower limit of detection corresponds to a particle size of 0.3 µm. The device operates according to the following principle (Figure 14.2.3). A volume in which the measurement is to be made is defined by purely optical means within a flow duct, by imaging a diaphragm. The particle size distribution is determined by measuring and analyzing pulses of scattered light. In this method it must be sufficiently probable that there is always only one particle in the measurement volume. The device has also been used successfully to investigate dispersing processes and the mechanisms where by various de-agglomeration mechanisms of various solid fractions de-agglomerate, and to determine drop size distributions produced by aerosol nozzles.

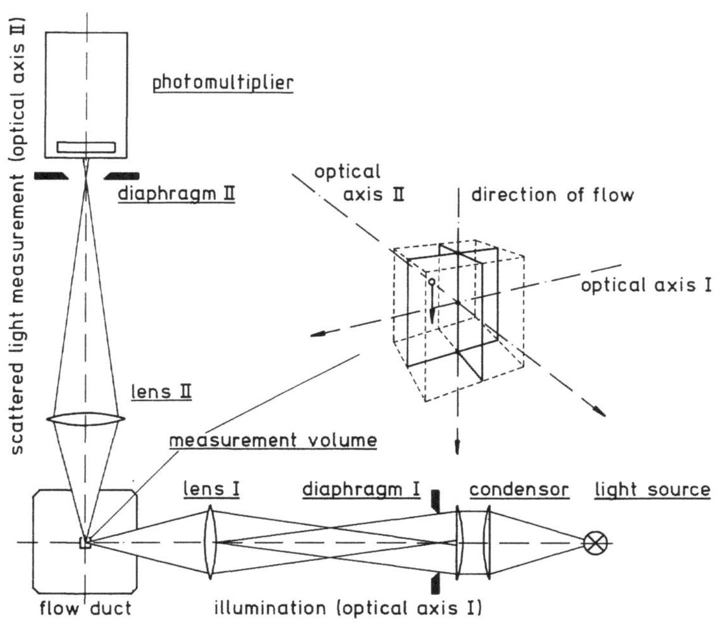

Figure 14.2.3 Schematic diagram of a scattered light measuring device

14.3 Devices for sampling granular and lumpy material

The simplest sampling methods are those which are carried out manually, by means of sampling scoops or sampling spoons. While sampling scoops are open at the top and on one side, and are hence suitable for scooping out stationary material, a sampling spoon is open only at the top and serves for taking samples from falling streams of granular material. Sampling forks are only suitable for lumpy materials, for example coke. Drive pipes and sampling corers have proved suitable for taking samples from railroad cars, ships and stores. Unlike scoops, they sample not just the surface but reach depths of several meters. In its simplest form, a drive pipe consists of a 2 - 4 mm thick steel tube having an internal diameter of 30 - 50 mm and a length of 1.50 m (in special cases up to 4 m). The drive pipe shown in Figure 14.3.1 is particularly suitable for free-flowing pulverulent materials. The pipe is forced into the material while open, is closed at the lower end by turning the handle and is then extracted.

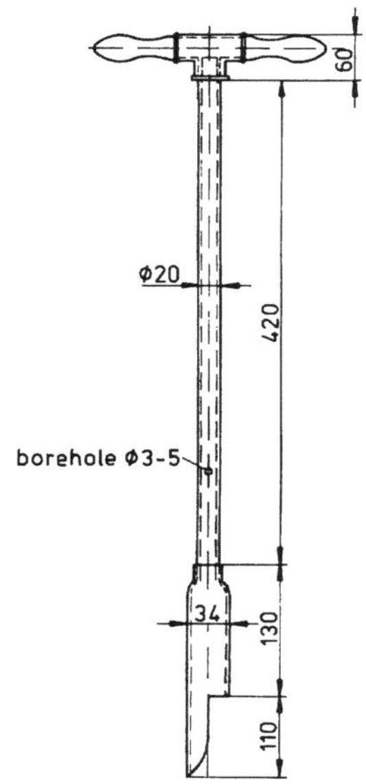

Figure 14.3.1 Drive pipe for free-flowing bulk material

The sampling corer is suitable for sticky materials, such as, for example, moist soils. It is made of two rigid curved pieces of sheet metal which are mounted on a crosspiece. The sampling corer is introduced into a heap of material by means of a corkscrew motion.

The best known sampling vessel for sampling free-flowing bulk material is the slot sampler. A slot sampler is normally moved at right angles trough a falling stream of material.

If the material to be investigated is being conveyed on a conveyor belt, a chute or a screw, manual sampling is being replaced to a steadily increasing extent by automatic devices. In the case of a conveyor belt, a simple method is provided by a stationary stripper at right angles to the conveying direction, which picks off an exact cross-section of the material. The stripper plate must be passed sufficiently closely to the surface of the conveyor belt to remove all the material[42] (Figure 14.3.2).

Pendulum samplers (Figure 14.3.3) and impact samplers (Figure 14.3.4) are used for automatic sampling at points where

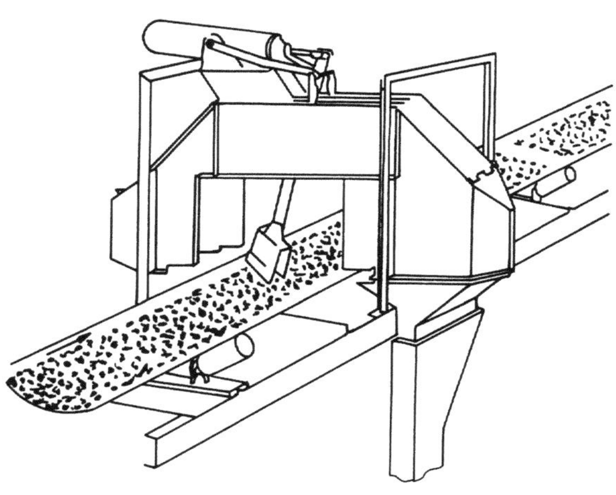

Figure 14.3.2 Sample stripper

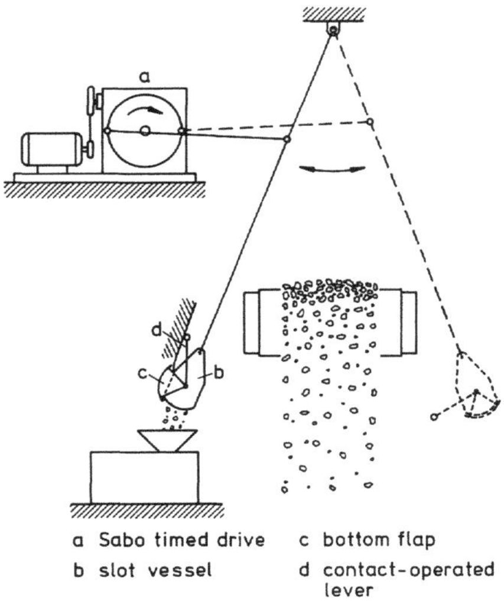

Figure 14.3.3 Pendulum sampler

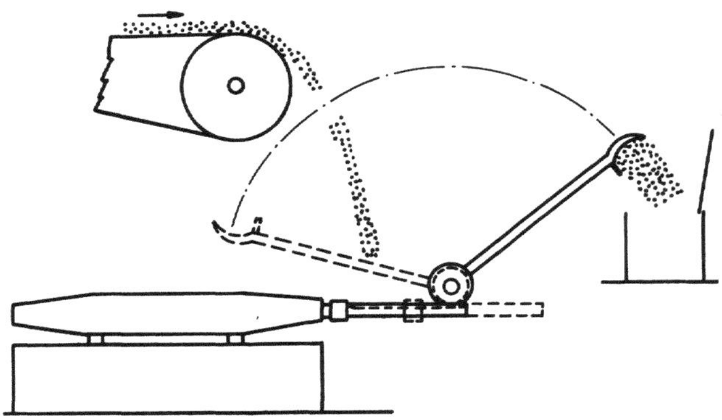

Figure 14.3.4 Impact sampler

material is discharged from a belt or chute. These devices take samples of the material at regular time intervals, either at right angles to the discharge direction, or in the discharge direction. The pendulum sampler consists of a slot vessel, suspended so that it can swing, which is passed through the material by means of a system of rods. On return into the rest position, the material is discharged by opening the bottom flap[43].

The impact sampler knocks material out of the falling stream of material, in the direction in which it is being conveyed, by means of a batten having a concave shape.

A simple slot device (Figure 14.3.5) serves for taking samples from a conveyor chute. A part stream is diverted, as a sample, from the center of the chute by means of a vertical slot. The part stream remains proportional to the conveyed stream as the depth of the layer of material on the chute and the conveying speed fluctuate.

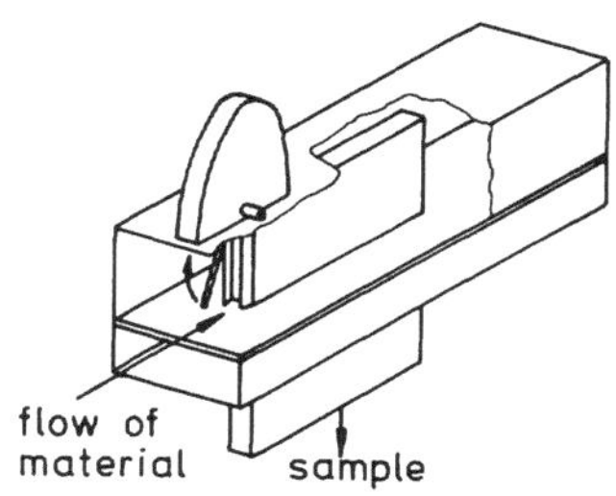

Figure 14.3.5 Slot device for taking samples from a conveyor chute

Free-flowing material can be diverted from hoppers and conveying pipes with the aid of a cellular-wheel sampler of a worm screw which has a number of slots and openings on its surface[44]. The sampling screw conveyor can also divert a part stream from a larger screw conveyor (Figure 14.3.6). If the speed of rotation of the sampling screw is constant, the rate at which material is removed for sampling is independent of the rate at which material is being conveyed by the large screw.

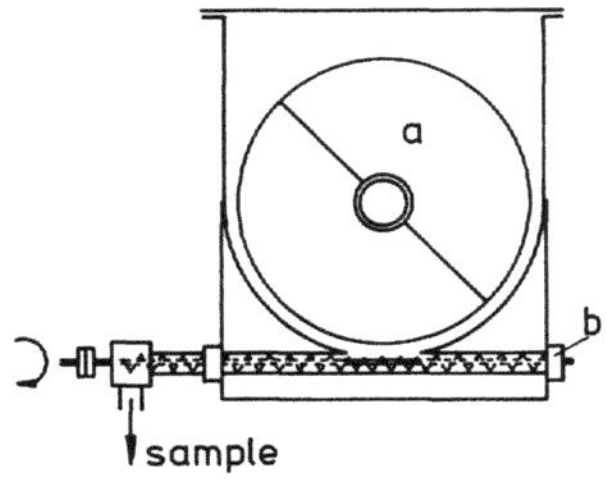

Figure 14.3.6 Sampling screw conveyor

14.4 Sample splitters

In addition to comminution and blending, sample splitting is an important step in the preparation of the sample. As in the case of sampling, splitting can be carried out manually or automatically. By hand, reduction is carried out by the so-called quartering method. To do this, the particulate material poured on to a flat surface so as to form a cone. On being added, fresh material should strike the point of the cone as centrally as possible. The cone is divided into four quadrants by means of two mutually perpendicular cuts. In each case the two opposite quadrants are combined and, if necessary, subjected to the same process until the sample size for analysis has been reached. It is advantageous to predetermine the split before pouring on the material, by means of a sheet metal cross.

In the Riffel splitter (Figure 14.4.1), the sample is poured from a suitable vessel, into a battery of about ten open chambers which are so arranged that any two adjacent chambers permit the material to flow out towards two different sides.

The material is recombined on each side. Each splitting step thus produces two new partial amounts. Again, splitting is carried out until the final sample size is reached.

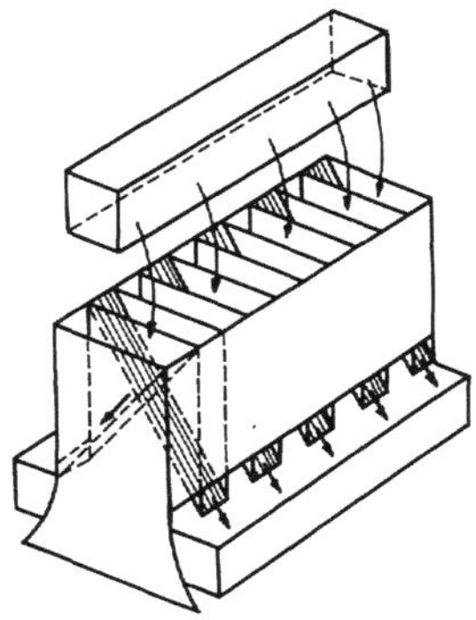

Figure 14.4.1 Riffle splitter

In a rotary sample splitter, a vibrating chute feeds a laboratory sample, as a slow-flowing stream of material, to a rotating system of cutters, whence it is distributed into individual vessels. The Retsch sample splitter (Figure 14.4.2) is operated by means of a feed device developed by Leschonski [20].

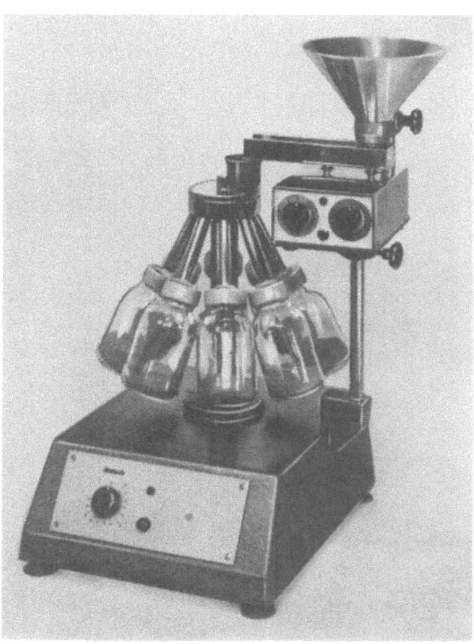

Figure 14.4.2
Rotary sample splitter

Rotating at a speed of 120 rpm, this feed device divides the material between 8 separate flasks. The rotary sample splitter is superior to all other samples-splitting methods, at least as far as free-flowing materials are concerned, as has been repeatedly confirmed[45,46].

The operating principles on which the Riffle splitter and the rotary sample-splitter are based are also made use of in large-scale and industrial equipment. However, despite the superiority of the rotary sample-splitter, the principle of the Riffle splitter is preferred for these larger units, because of its simplicity and because the units can easily be connected in series.

Notes

1) In principle, gases and liquids also consist of "particles", the molecules or atoms, but the sample size is so much larger compared with the "particles" that the size of the molecules does not have any influence on the sampling error.

2) Kolmogoroff, A.N., Grundbegriffe der Wahrscheinlichkeitsrechnung, Springer, Berlin 1933

3) Fisz, M., Wahrscheinlichkeitsrechnung und mathematische Statistik, VEB Deutscher Verlag der Wissenschaften, Berlin 1965

4) B.E. Netto, Lehrbuch der Kombinatorik, Teubner-Verlag, Berlin, 1927

5) Kreyszig, E., Statistische Methoden und ihre Anwendungen, Vandenhoeck u. Reprecht, Göttingen 1968

6) For the proof see, for example, Kreyszig, E., Statistische Methoden und ihre Anwendungen, Vandenhoeck + Ruprecht, Göttingen, 1968, page 171

7) Linder, A. Statistische Methoden, Birkhäuser Verlag, Basel 1960

8) Graf, Henning, Stange, Formeln und Tabellen der mathematischen Statistik, Springer-Verlag, Berlin, Heidelberg, New York 1966, or Sachs, L., Statistische Auswertemethoden, Springer-Verlag, Berlin, Heidelberg, New York 1969

9) Stange, K., Angewandte Statistik, Part 2, Springer-Verlag, Berlin, Heidelberg, New York 1971, page 408

10) Wilrich, P.-Th., Statistische Gesichtspunkte bei der Probenahme von körnigen Massengütern, Prüfsiebung, Siebtechnik GmbH, 1969

11) Stange, K., Die Mischgüte einer Zufallsmischung als Grundlage zur Beurteilung von Mischversuchen (The degree of mixing of a random mixture as the basis for assessing mixing trials), Chem.-Ing.-Techn. 26 (1954), 331/337

12) Linder, A., Statistische Methoden, Birkhäuser Verlag Basel, 1960, page 365 et seq.

13) Lacey, M.L. The mixing of solids particles, Trans. Inst. Chem. Engrs. 21 (1943) 53

14) Gy, P., Echantillonnage pour une analyse granulometrique, (Sampling for particle-size analysis) Annales des Mines 145 (1956) 4, 37/42

15) Poole, K.R., Taylor, R.F., Wall, G.P., Mixing of powders to fine scale homogeneity, Trans. Inst. Chem. Engrs. 42 (1964) P 305

16) Kristensen, H.G., Statistical properties of random and non-random mixtures of dry solids, Powder Techn. 7 (1973), 249/257

17) Koglin, B., Unveröffentlichte Messergebnisse (Unpublished experimental results), Inst. für Mech. Verfahrenstechnik Institute for Mechanical Process Technology , Karlsruhe University.

18) Pawlowski, J., Zur Statistik der Homogenisierungsprozesse (A contribution to the statistics of homogenization processes), Chem.-Ing.-Techn. 36 (1964) 11, 1089/1098

19) Rumpf, H. Mechanische Verfahrenstechnik, Carl Hanser Verlag, München, Wien, 1975, P.8 sequ.

20) Koglin,B., Leschonski, K., Alex, W., Teilchengrössenanalyse T 2: Probenahme Particle-Size Analysis T 2: Sampling, Chem.-Ing.-Techn. 46 (1974), 289/292

21) Tanaka, T. Prefential grinding mechanism of binary mixtures, Symposion Zerkleinern, VDI-Verlag, Düsseldorf 1962, 360

22) Kendall, M.G., Moran, P.A., Geometrical Probability, Griffen et Co., London 1963, P.86

23) Wicksell, S.D., The corpuscle problem, Biometrika 17 (1925) 84/99

24) Gupte, A., Experimentelle Untersuchung der Einflüsse von Porosität und Korngrößenverteilung im Widerstandsgesetz der Porenströmung. Dissertation Universität Karlsruhe 1970

25) Rumpf, H., Vorlesung "Mischen" Universität Karlsruhe SS 1964

26) Debbas, S., Über die Zufallsstruktur von Packungen aus kugeligen oder unregelmäßig geformten Körnern. Dissertation Universitaät Karlsruhe 1965

27) Krumbein, W.C., Thin section mechanical analysis of indurated sediments, J. Geology 43 (1935) 482/496

28) Courant, R., Hilbert, D., Methoden der Mathematischen Physik I, Springer-Verlag 1931, P134

29) Mihira, K., Okasawa, T., Nakayama, A., Geometry of polymeric form or cellular structures (I), Kolloid-Z. and Z. f. Polymere 222 (1968), 135 et seq.

30) Kausch von Schmeling, H.H., Tschoegl, N.W., On the transformation of histograms with geometrically progressing class intervals. J. of Composite Materials 3 (1969) 6, 445/452

31) Müller, W., Untersuchungen über Mischzeit, Mischgüte und Arbeitsbedarf in Mischtrommeln mit rotierenden Mischelementen, Dissertation Karlsruhe 1966

32) Lacey, P.M., Mirza, F.S., A Study of the Structure of Imperfect Mixture Parts I ans II, Powder Technology 14 (1976), 17/24, 25/33

33) Stange, K., Probenahme vom Band, Metrika 1(1958) 177/222

34) Analyse d. Metalle, (Analysis of Metals), Volume 3, Probenahme Sampling Springer Verlag, Berlin, Heidelberg, New York 1975

35) Passavant-Werke Michelbacher Hütte, Germany

36) Despora Milton Ray GmbH, Neu-Isenburg, Germany

37) Drägerwerk AG Lübeck, Germany

38) Batel, W., Einführung in die Korngrößenmeßtechnik, Springer Verlag, Berlin Heidelberg New York, 1971

39) Vitols, V., Theoretical limits of Errors Due to Anisokinetic Sampling of particulate Matter, J. of Air Pollution 16 (1966)2, 79/83

40) Rüping, G., Die Bedeutung der geschwindigkeitsgleichen Absaugung bei der Staubmessung mittels Entnahmesonden, Staub 28(1968)4, 137/144

41) Umhauer, H., Ermittlung von Partikelgrößenverteilung in Aerosolströmungen hoher Konzentration mit Hilfe einer Streulichtmeßeinrichtung, Chem.-Ing.-Techn. 47(1975)7, P.297

42) Köhlung, R., Entwicklungstendenzen bei der Probenahme und Probevorbereitung fester mineralischer Rohstoffe, Aufbereitungstechnik (1972)12, 698/704

43) Block, E., Einige Beispiele der automatischen Probenahme, Glückauf 101(1965)4, 255/264

44) Hilbig, M., Probenahme und Probeaufbereitung von festen und grobkörnigen Gütern, Aufbereitungstechnik (1972)11, 705/712

45) Kaye, B.H., Dissertation, London 1962

46) Allen, T., Khan, A., Critical Evaluation of powder sampling procedures, The Chem. Eng. 1970, CE 108/CE 112

Literature

A. Text-books and monographs

I. Statistics

Bicking, Ch.A. — The Sampling of Bulk Materials
Material Research + Standards 7(1967)3,95/116

Cramer, H. — Mathematical Methods of Statistics. 8th Princeton: Princeton University Press, 1958.

Duncan, A.J. — Bulk Sampling: Problems and Lines of Attack
Technometrics 4 (1962) 2, 319/344

Fisz, M. — Wahrscheinlichkeitsrechnung und Mathematische Statistik
VEB Deutscher Verlag der Wissenschaften,Berlin 1965

Hald, A. — Bayesian Single Sampling Attribute Plans for Continuous Prior Distributions.
Technometrics 10 (1968), 667 - 683

Hald, A. — The Compound Hypergeometric Distribution and a System of Single Sampling Inspection Plans Based on Prior Distributions and Costs.
Technometrics 2 (1960), 275 - 340

Herdan, G. — Small particle statistics
Butterworths London 1960

Koch, W. u.a. — Probenahme von Massengütern und ihre Bedeutung für die Hüttenwerke
Thyssenforschung 2 (1970) 1, 6/11

Kolmogoroff, A. — Grundbegriffe der Wahrscheinlichkeitsrechnung
Springer Verlag Berlin 1933

Lindley, D.V. — Introduction to Probability and Statistics from Bayesian Viewpoint.
Part 1: Probability.
Part 2: Inference.
Cambridge: University Press 1965

Neumann, J.v., Morgenstern, O. — Theory of Games and Economic Behavior.
3rd. ed. Princeton, N.J.: Princeton University Press, 1953

Schmetterer, L. — Einführung in die mathematische Statistik
Springer Verlag Wien 1966

Schmitt, S.A.	Measuring Uncertainty. An Elementary Introduction to Bayesian Statistics. Reading, Mass.: Addison-Wesley 1969
Stange, K.	Angewandte Statistik 1. Teil Eindimensionale Probleme 2. Teil Mehrdimensionale Probleme Springer Verlag, Berlin, Heidelberg, New York 1970/71
van der Waerden,	Mathematische Statistik Springer Verlag, Berlin, Heidelberg, New York 1965
Wilks, S.S.	Mathematical Statistics New York: Wiley, 1962

II. Sampling

Cochran, W.G.	Sampling techniques John Wiley and Sons, Inc. New York (1967)
Cochran, W.G.	Stichprobenverfahren Walter de Cruyter, Berlin, New York (1972)
Dalenius, T.	Sampling in Sweden, Contributions to the Methods and Theories of Sample Survey Practice Almquist + Wiksell, Stockholm, 1957
Deming, W.E.	Some Theory of sampling John Wiley and Sons, Inc., New York 1961
Gy, P.	Theorie et Pratique de l Echantillonnage des Matieres Morcelles Edition PG, Cannes 1975
Gy, P.	Sampling of Particulate Materials Elsevier Scientific Publishing Company Amsterdam, Oxford, New York 1982
Handbuch	Probenahme - Analyse der Metalle, Band 3, Springer Verlag, Berlin, Heidelberg, New York
Kellerer, H.	Theorie und Technik des Stichprobenverfahrens Einzelschriften der Deutschen Statistischen Gesellschaft München 1953
Smith/Duncan	Sampling Statistics and Applications Mc Graw-Hill Book + Co, New York, London 1945
Smith, R., James, G.V.	The Sampling of Bulk Materials The Royal Society of chemestry,London 1981

Sommer, O. Probenahme, Probemenge, Probeverarbeitung
Staub (1955) 42, 644/677

Stuart, A. Basic ideas of Scientific Sampling
Charles Griffin + Co. London 1968

Tanner, L., Some Problems in the Sampling of Bulk Materials
Am. Soc. Testing Mats Proceedings
49 (1949), S. 1181

B. Papers

Distributions by Sampling (chapter 3 and 4)

Benedict, R.P. Analysis of Experimental Data
Instr. + Control Systems 42 (1969) 6, 125/131

Cramer, H. Über eine Eigenschaft der normalen Verteilungsfunktion
Math. Zeitschrift 41 (1936)

Hajek, J. Limiting distributions in single random sampling from a finite population
Publ. math. Inst. Hungar. Acad. Sci. Ser.
A 5 (1960), 361/371

Hald, A. The Mixed Binomial Distribution and the Posterior Distribution of p for a Continuous Prior Distribution
J.R. Stat. Soc. B 30 (1968), 359/367

Shindle, Z., Gallily, I. On the Applicability of the Poisson Distribution to Suspension Spray-Drying
J. Colloid + Interface Sci. 52 (1975) 3, 538/542

Stange, K. Über die Probenahme bei Massengütern
Qualitätskontrolle 4 (1959), 25/28

Stange, K. Stichprobenverfahren bei Beurteilung von Massengütern, insbesondere von Kohle
Mitt.blatt für Math. Statistik
6 (1954) 3, 204/220

Subrahmanya, M.T. On Generalized Estimators in Sampling from Finite Populations
Fresenius Z. analyt. chemie 222 (1966) 2, 234/239

Visman, J. Test on the Binomial Sampling theory for Heterogeneous coal
Amer. Soc. Test. Mats., S.T.P. No 162, 1954, 141/152

Confidence intervals (chapter 5)

Box, G.E.P., Tiao, E.G. Bayesian Inference and Statistical Analysis
Reading, Mass.: Addison-Wesley 1973

Bunday, B.D. Methods for Population Estimation by Random Sampling
Powder Technology 12 (1975) 283/286

Gosset, W.S. The probable error of a mean
Biometrica 6 (1908) 1/25

Graf, U., Henning, H.J. Mathematisch-statistische Grundlagen bei der Probenahme und Probewertung von Erzen, Metallen und Rückständen
Z. Erzbergbau Metallhüttenwes. 5 (1952) 127/131

Hald, A. Bayesian Single Sampling Attribute Plans for Discrete Prior Distributions.
Mat. Fys. Skr. Dan. Vid. Selsk. 3 (1965) 1/88

Hald, A. Asymptotic Properties of Bayesian Single Sampling
Plans. J.R. Stat. Soc. B 29 (1967) 162/173

Harris, C.C. Estimating Small Populations by Sampling
Powder Technology 12 (1975) 1, 85/91

Kiessewetter, F. Über die Anwendung statistischer Methoden der Qualitätskontrolle und der Probenahme basischer, feuerfester Erzeugnisse
Materialprüfung 6 (1964) 11, 395/398

Kortlandt, D., Zwart, R.L. Find optimum timing for sampling product quality
Chem.Engng. 78 (1971) 25, 66/72

Stange, K. Die Berechnung von einseitigen Toleranzgrenzen aus Mittelwert und Standardabweichung einer Probe
Qualitätskontrolle 3 (1958) 7/8

Stange, K. Ein Näherungsverfahren zur Berechnung optimaler Pläne für messende Prüfung bei bekannter Verteilung der Schlechtanteile in den vorgelegten Liefermengen
Metrika 10 (1966) 92/136

Wenzel, H. Der Einfluß der Stichprobenentnahme und des Stichprobenumfangs auf die Aussagekraft von Messungen
VDI-Z. 108 (1966) 35, 1717/1720

Sampling Procedures (chapter 6)

Becker, R.M. Some generalized probability distributions with special reference to the mineral industries US Bureau of Mines RI 6329 : 53 pp, RI 6652 : 101 pp, RI 6598 : 79 pp, RI 6627 : 57 pp, RI 6768 : 60 pp (1964-66)

British Standards Institution The sampling of coal and coke
BS 1017 Part 1 : Coal : 124 pp.
Part 2 : Coke :106 pp. (1960)

Brunton, D.W. The theory and practice of ore sampling
Trans. AIME, 25 : 826 (1895)

Geary, R.C. The frequency distribution of the quotient of two normal variables
J. Roy. Stat. Soc. 93 : 442 (1930)

Ghosh, S.P. Optimum Allocation in Stratified Sampling with Replacement
Fresenius Z. analyt. Chemie 222 (1966) 2, 212/221

Gy, P. Sampling - The error committed in size distribution
Jamshedpur Congress. Ind. Min. J. (1957)

Gy, P. A new Theory of ore sampling
AIME annual meeting, New-Orleans (1957)

Gy, P. Le principe d'equiprobabilite. Ann. Min. (Dec. 1964) 779/794

Gy, P. Sampling of ores and metallurgical products during continuous transport
Trans. IMM 74 (1965) 165/199

Harnby, N. The statistical analysis of particulate mixtures
Powder technol. 5 (1971/72) 81/85

John, B. Statistische Aspekte bei der Probenahme von Schüttgütern
Zement-Kalk-Gips (1970) 1, 32/39

Manning, A.B. The Use of "Random Numbers" in the Theory of Coal Sampling
J. Inst. Fuel 26 (1953), 107

Matheron, G.	The intrinsic random functions and their applications Adv. in Appl. Prob. 5 (1973) 439/468
Paul, H.	Zusammenhänge zwischen der Probenahme von Steinkohlen und der Genauigkeit des ermittelten Aschegehaltes Glückauf 94 (1958) 27/28, 907/918
Stange, K.	Die wirtschaftliche Verteilung des Prüfaufwandes auf die zwei Stufen eines Modells mit unterschiedlicher Zahl und Varianz und unterschiedlichen Erhebungskosten der Einheiten zweiter Stufe Metrika 9 (1965) 195/211
Stange, K.	Statistische Methode für den Verfahrensing. Chemie-Ing.-Techn. 32 (1960) 3, 143/154
Wilson, A.D.	The sampling of silicate rock powders for chemical analysis Analyst 89 (1964) 1054, 18/30

Sampling from a random mixture (chapter 7)

Buslik, D.	Mixing and Sampling with special Reference to Multi sized Granular Material Amer. Soc. Testing Mats. Bull. No. 165 (1950), S. 66
Gy, P.	Echantillonnage pour une analyse granulometrique Annales des Mines 145 (1956) 4, 37/42
Harnby, N.	The estimation of the variance of samples with drawn from a random mixture of Multisized particles The Chem. Eng. (1967) 12, CE 270/CE 271
Hutschenreiter, W.	Fehlerfortpflanzung bei der Probenahme Bergakademie 17 (1965) 9, 537/542
Kristensen, H.G.	Statistical properties of random and nonrandom mixtures of dry solids (Pt1 and Pt2) Powder Technol. 7 (1973) 5, 249/257 8 (1973) 3/4, 149/157
Lacey, M.C.	The mixing of solids particles Trans.Inst. Chem. Engrs. 21 (1943) 53
Marc de Chaza, L.E. Hung, C.	A study of the Effect of Sample Size on the Analysis of Powder Mixtures Amer. Inst. Chem. Engrs. 14(1968)1, 169/173

Poole, K.R., Tayler, R.F., Wall, G.P. — Mixing of powders to fine scale homogeneity
Trans. Inst. Chem. Engrs. 42 (1964) T 305

Pawlowski, J. — Zur Statistik der Homogenisierprozesse
Chemie-Ing.-Techn. 36 (1964) 11, 1089/1098

Ries, H.B. — Mischgüte, Problematik, Prüfmethoden und Ergebnisse
Aufbereitungstechnik (1976) 1, 16/28

Rumpf, H. — Mechanische Verfahrenstechnik
Carl Hanser Verlag München, Wien 1975

Sommer, K. — Physikalische Vorgänge beim Conchieren
Dissertation Karlsruhe 1974

Sommer, K. Rumpf, H. — Varianz der stochastischen Homogenität bei Körnermischungen und Suspensionen und praktische Ermittlung der Mischgüte
Chemie-Ing.-Techn. 46 (1974) 257 ff.

Stange, K. — Zur Beurteilung der Güte einer Mischung aus körnigen Stoffen bei bekannten Siebdurchgangslinien der Komponenten
Chemie-Ing.-Techn. 36 (1964) 3, 296/302

Stange, K. — Beurteilung von Mischgeräten mit Hilfe statistischer Verfahren
Chemie-Ing.-Techn. (1954) 3, 150/155

Stange, K. — Die Mischgüte einer Zufallsmischung aus drei und mehr Komponenten
Chemie-Ing.-Techn. 35 (1963), 580/582

Stange, K. — Die Mischgüte einer Zufallsmischung als Grundlage zur Beurteilung von Mischversuchen
Chemie-Ing.-Techn. 26 (1954), 331/337

Vollbrecht, H. — Zur Frage der Mischgüte bei körnigen Stoffen
CZ-Chemie-Technik 1 (1972) 3, 109/112

<u>Sampling for grani-size analysis of a particle distribution (chapter 9)</u>

Göll, G., Helfricht, R. — Zum Problem der Probenahme für Korngrößenanalysen
Freiberger Forschungsh. A 531, Leipzig 1975, 239/255

Guitard. P., Sucker, H.	Beitrag zur Bestimmung der wahren Volumenverteilungssumme von Haufwerken mit Hilfe von automatischen Teilchenzählgeräten Pharm. Ind. 36 (1974) 5, 321/327
Guitard, P., Leuenberger, H., Sucker, H.	Maximal zulässige Partikelgröße von Hilfsstoffen in Mischungen für feste Arzneiformen Pharm. Ind. 36 (1974) 4, 253/257
Guitard, P., Rees, J.	Maximale zulässige Partikelgrößenverteilung von Wirkstoffen für feste Arzneiformen in niederer Dosierung Pharm. Ind. 36 (1974) 4, 253/257
Johne, R., Ramanujam, M.	Genauigkeit der Kornanalyse mit dem Photosedimentometer Staub 23 (1963) 5, 265/269
Koglin, B., Leschonski, K., Alex, W.	Teilchengrößenanalyse. Teil 2: Probenahme Chemie-Ing.-Techn. 46 (1974) 7, 289/292
Lai, F.S., Wang, R.H., Fan, L.T.	An application of nonparametric statistics to the sampling in solids mixing Powder Technol. 10 (1974) 1/2, 13/21
Lines, R.W.	Some observation on sampling for particle size analysis with the coulter-counter Powder Technol. 7 (1973) 3, 129/136
Mayne, F.	Die Verfahren der Probenahme zwecks Korngrößenanalyse von Pulvern Sci. techn. Pharm. 4 (1975) 2, 247/256
Rumpf, H.	Mechanische Verfahrenstechnik Carl Hanser Verlag, München, Wien 1975
Rumpf, H.	Über die Feinheitsbestimmung von technischen Stäuben Staub 25 (1965) 1, 15/22
Schneider, U.	Stoffbezogene mathematisch-statistische Untersuchungen über Korngrößenverteilungen fein aufgemahlener Minerale Aufbereitungstechn. 15 (1974) 10, 535/541
Sommer, K.	Einfluß des Probevolumens auf den mittleren Fehler der Korngrößenanalyse Chem.-Ing.-Techn. 47 (1975) 3, 95/96
Walkenhorst, W.	Ergebnisse mit einem neuen Verfahren zur Bestimmung der Teilchengrößenverteilung von Schwebestaub Staub-Reinhalt. Luft 30 (1970) 11, 458/464

Sampling error when sampling from ores and fuels (chapter 10)

Bertholf, W.M. — The Development of the theoretical basis of Coal Sampling
Amer. Soc. Testing Materials, S.T.P. 162, 1954, 5/34

Gy, P. — The sampling of broken ores - A review of principles and practice
In : The Institution of Mining and Metallurgy, London (Editor), Geological, Mining and Metallurgical Sampling (1973) 194/205 and discussion 261/263

Melkikh, V.I. — Statistical Interpretation of Equally-Probable chemical sampling of ores and Benefication Products
Industrial Laboratory 41(1975)7, 1068/1071

Pethö, S. — A szemcseterfogattal aranyos valoszinüsegü mintavetelröl
Müszaki Pudomany (1973) 227/257

Richards, R.H. — Ore dressing
(1908) Vol. 2 843/852, Vol. 3 1571/1578, Vol. 4 2031/2033, Mac-Graw Hill, New York

Random packings (chapter 11)

Bach, C. — Über die Bestimmung von charakteristischen Größen einer Kugelverteilung aus der unvollständigen Verteilung der Schnittkreise
Metrika 9 (1965) 228/233

Blum, E.H., Wilhelm, R.H. — A statistical Geometric Approach to Random-packed beds
A.I.Ch.E.-I.Chem.E. Sympos. Ser. 4 (1965) 21/27

Bridgwater, J., Scott, A.M. — Statistical models of packing: Application to gas absorption and solids mixing
Trans. Instn.chem.Engrs. 52(1974)4, 317/324

Debbas, S. — Über die Zufallsstruktur von Packungen aus kugeligen oder unregelmäßig geformten Körnern Dissertation Karlsruhe 1965

Debbas, S., Rumpf, H. — On randomness of beds packed with spheres or irregular shaped particles
Chem.Eng.Science 21 (1966) 583/607

Eckhoff, R.K. Enstad, G.	Derivation of Particle Sizes Distributions from Measurements on Plane Sections through Particle Beds. Effect of Regularity of Particle Packing Powder Technology 11 (1975) 1/10
Gupte, A.	Experimentelle Untersuchung der Einflüsse von Porosität und Korngrößenverteilung im Widerstandsgesetz der Porenströmung Dissertation Universität Karlsruhe 1970
Kausch v.Schmeling Tschoegl, N.N.	On the transformation of Histogramms with Geometrically Progressing Class Intervalls J. of Compsite Materials 3(1969)6, 445/452
Kendall/Moran	Geometrical Probability Charles Griffin + Co., London 1963
Krumbein, W.C.	Thin section mechanical of industated sediments J. Geology 43 (1935) 482/496
Mihira, K., u.a.	Geometry of Polymeric Foam or Cellular Structures (I) Kolloid-Zeitschr. + Zeitschr. f. Polymere 222 (1968), 135/140
Montgomery, D.W.	Der Zeiss Teilchengrößenanalysator bei der Entwicklung künstlicher Latex-Polymere Zeiss-Werkzeitschrift 10 (1962) 44, 57/59
Rodgers, C.A.	The packing of equal spheres Proc. London Math. Soc. 8 (1958) 3, 609/620
Rose, H.E.	The density of packing of two component powder mixtures Powder Metallurgy 8 (1965) 15, 20/37
Rumpf, H., Debbas, S., Schönert, K.	Über die Struktur von Packungen körniger Stoffe Chem.-Ing.-Techn. 39 (1967) 3, 116/125
Schwanda, F.	Der Hohlraumgehalt von Korngemischen Beton (1959) 1, 12/17
Wicksell, S.O.	The corpuscle problem Biometrica 17 (1925) 84/99

Non-random mixtures (chapter 12)

Bourne, J.R.	Variance sample size relationships for complete mixtures Chem.Eng. Science 22 (1967) 693/700

Bourne, J.R.	Some statistical relationships for powder mixtures Trans.Instn.Chem.Engrs. 43 (1965) 191, CE 198/200
Hutschenreiter, W.	Fehlerrechnung und Optimierung bei der Probenahme Freiberger Forschungsh. A 531, Leipzig 1975, 231/238
Lacey, P., Mirza, F.	A study of the Structure of Imperfect Mixtures of Particles P. I + II Powder Technology 14 (1976) 17/33
Müller, W.	Untersuchungen über Mischzeit, Mischgüte und Arbeitsbedarf in Mischtrommeln mit rotierenden Mischelementen Diss. Universität Karlsruhe 1966
Ring, R.J., Buchanan, R.H. Doig, I.D.	The discharge of granular material from hoppers submerged in water Powder Technol. 8 (1973) 117/125
Stange, K.	Genauigkeit der Probenahme bei Mischungen körniger Stoffe Chem.-Ing.-Techn. 39 (1967) 9/10, 585/592

Sampling from a conveyer belt (chapter 13)

Stange, K.	Probenahme vom Band Metrika 1 (1958) 3, 177/222

Devices for sampling liquids (chapter 14.1)

Anonym	Sampling from a Vessel under Vacuum Brit.chem.Engng. 3 (1958) 6, 335
Anonym	Small-Volume Composite Sampler Developed Chem.Engng. News 41 (1963) 28, 61
Burchard, C.H.	Automatische Entnahme wassermengenabhängiger Abwassermischprobe Das Gas- und Wasserfach 107 (1966) 44, 1248/1251
Calcote, R.E.	Siphon Arrangement Takes Continuous Samples Chem.Engng. 63 (1956) 7, 214
Schwarz, D., Doken, P.C.	Proportional Sampler is Automatic Chem. Engng. 66 (1959) 6, 184

Devices for sampling gases (chapter 14.2)

Anonym	Sampling Valve Survey Instr. Controll. Syst. 43 (1970) 1, 89/95
Biheller, J., Resch, W.	Erprobung einer Sammel- und Meßmethode für die routinemäßige Bestimmung von Fluor in der bodennahen Atmosphäre Staub-Reinhalt. Luft 31 (1971) 1, 9/11
Chipman, J.C., Massey, M.T.	Proportional Sampling System for the Collection of an Integrated Auto Exhaust Gas Sample J. Air Poll, Contr. Ass. 10 (1960) 1, 60/69
Higashino, I.	A New Electromagnetic Gas Sampling Valve Bull. of JSME 4 (1961) 15, 591/598
Lenze, B.	Probeentnahme und Analyse von Flammengasen Chem.-Ing.-Techn. 42 (1970) 5, 287/292
Lodge, J.P., Jr.	The Use of Hypodermic Needles as Critical Orifices in Air Sampling J. Air Pollut. Contr. Ass. 16 (1966) 4, 197/200
Natelson, S., Stellate, R.L.	Apparatus for Extraction of Gases for Injection into the Gas chromatograph Analytic. Chem. 35 (1963) 7, 847/851
Otte, E.	Autosampler AS 41. Ein automatischer Probengeber für die Gas-chromatographie Chromatogr. 4 (1971) 12, 588/592
Pack, M.R., Hill, A.C., Benedict, H.M.	Sampling Atmospheric Fluorides with Glass Fiber Filters J. Air Pollut. Contr. Ass. 13 (1963) 8, 374/377
Parker, W.R., Huey, N.A.	Multipurpose sequential samplers J. Air Poll. Contr. Ass. 17(1967)6, 388/391
Rogers, B.L. Prausnitz, J.M.	Sample-Extrusion Apparatus for High-Pressure Vapor-liquid Equilibria I + EC Fundam. 9 (1970) 1, 174/177
Rohrschneider, L., Jaeschke, A., Kubik, W.	Untersuchung der Luftverunreinigungen in Höhen bis zu 500 m über Industriegelände Chem.-Ing.-Techn. 43 (1971) 18, 1010/1017
Ross, St. S.	Designing your plant for easier emission testing Chem. Engng. 79 (1972) 14, 112/116, 118

Devices for sampling dusts (chapter 14.2)

Achinger, W.C., Shigehara, R.T.	A Guide for Selected Sampling Methods for Different Source Conditions J. Air Poll, Contr. Ass. 18 (1968) 605/609
Avy, A.P., Benarie, M., Hartagensis, F.	Vergleich von Probenahmegeräten und -verfahren für Stäube Staub-Reinhalt. Luft 27 (1967) 11, 469/480
Bartak, Jo.	Staubprobenahme in strömenden Gasen, zweidimensionaler Fall Staub-Reinhalt. Luft 34 (1974) 8, 295/300
Binek, B.	Probenahme feindisperser Aerosole für ein elektronenmikroskopisches Kornanalysenverfahren Staub 25 (1965) 7, 261/265
Breuer, H.	Experimentelle Untersuchungen über den Einfluß der Strömungs- und Ansauggeschwindigkeit auf die Meßanzeige verschiedener Staubmeßgeräte im Steinkohlenbergbau Staub 22 (1962) 3, 91/95
Chojnowski, B.	A Multipoint Solid and Gas Sampling Probe J. of the Institute of Fuel 37 (1964) 79/81
Coenen, W.	Ein neues Meßverfahren zur Beurteilung fibrogener Stäube am Arbeitsplatz Staub-Reinhalt. Luft 33 (1973) 3, 99/103
Conner, W.D.	An Inertial-Type Particle Separator for Collecting Large Samples J. Air Pollut. Contr. Ass. 16(1966)1, 35/38
Davies, C.N.	Deposition und Rentention von Staub in den menschlichen Atemwegen Staub 24 (1964) 8, 316
Dennis, R., Samples, W.R., Anderson, D.M., Silverman, L.	Isokinetic Sampling Probes Ind. Engng. Chem. 49 (1957) 2, 294/302
Drehmann, L.	Orifices Govern Inexpensive Sampler Chem. Engng. 66 (1959) 8, 186
Dunmore, J.H.,	An instrument for the sampling of respirable dust for subsequent gravimetric assessment J. Sci. Instrument 41 (1964) 669/672
Gisclard, J.Br.	The Compact Sequential Autom. Air Sampler J. Air Poll. Contr. Ass. 12 (1962) 2, 81/82

Hawksley, P.G.W.	Improved sampling equipment for solids in flue gases J. Inst. Fuel 31 (1958) 207, 147/160
Hemon, G., Pannetier, R., Philippot, J.C.	Aufbereitung von Meßproben atmosphärischer Aerosole im Hinblick auf ihre Auswertung und auf die systematische Überwachung der Radioaktivität Staub-Reinhalt. Luft 33 (1973) 10, 401/405
Houser, E.A.	Keeping Sample Lines Hot Control Engng. 7 (1960) 6, 123/125
Killingmo, O.H.	Kontinuierliche Meßverfahren f. Abgase Chem.-Ing.-Techn. 45 (1973) 1

Devices for sampling granular and lumpy material (chapter 14.3)

Anonym	Product Sampler for Fluid Bed Reactor Brit. chem. Engng. 8 (1963) 5, 353
Batel, W.	Einführung in die Korngrößenmeßtechnik Springer-Verlag Berlin, Heidelberg, New York
Blatter, A.O.	A Test on slotted revolving cylinder coal sampler Amer. Soc. Test. Mats. S.T.P. 162, (1954) 57/71
Blümel, G.	Automatisierung der Probenahme bei der Verladung Freiberger Forschungsh. A 316 (1964) 135/139
N. Eerste Nederlandse Cement Industrie (Enci)	Die Probeaufnahme von feinkörnigem Gut mit dem Enci-Probenehmer Aufbereitungstechn. 13 (1972) 11, 740/741
Emceka Gompper Köln-Heumar	Probenehmer für die Zementindustrie-Laboreinrichtung Aufbereitungstechn. (1974) 4, 218, 462
Hawes, R.W.M., Müller, L.D.	A Small Rotary Sampler and Preliminary Study of its Use A.E.R.E. - R 3051 (Harwell: UKAEA) 1960
Holland, W.D. Conway, R.E.	Three multi-stage stack samplers Chem. Engng. Progr. 69 (1973) 6, 93/95
Kennedy, W.R.	Powder Sampler Brit. Chem. Engng. 4 (1959) 2, 105
Meritt, A.D., Golden, J.T.	Low-Cost Continuous Sampler for plant Effluents Chem. Engng. 78 (1971) 7, 122

Mesters, J.	Die Enci Probenehmer für feinkörniges Gut Zement-Kalk-Gips 25 (1972) 3, 121/122
Visman, J.	The Importance of Automatic Sampling-Potentialities and Applications Fuel 29 (1950) 101/105
Allen, T., Khan, A.A.	Critical Evaluation of Powder Sampling Procedures Trans. Instn. Chem. Engrs. (1970) 238, CE 108/112
Charlier, R., Gossens, W.	Sampling a heterogeneous powder using a spinning riffler Powder technol. 4 (1971) 6, 351/359
Huntington, F.R.	Sampling systems a design approach Rock Products 78(1975)5, 118/120, 130/131

Sample splitters (chapter 14.4.)

Leonard, E.F.	Sampler Handles solids-Carrying streams Chem. Engng. 63 (1956) 12, 218/220
Maguire, Br. A., Harriss, G.W., Saunders, E.	Respirable - Dust Sampling in British Quarries Staub-Reinhalt. Luft 35 (1975) 11, 397/400
Meyer, P.B.	Staubmessungen in der holländischen keramischen Industrie Staub-Reinhalt. Luft 30 (1970) 10, 412/415
Morrow, N.L., Brief, R.S., Bertrand, R.R	Sampling and analyzing air pollution source Chem. Engng. 79 (1972) 2, 84/98
Fa. Du Pont	Analyzer Simplifies Difficult Sampling Problems Chem. Engng. 69 (1962) 18, 98/100
Popa, B., Lancan, V.	Die Wahrscheinlichkeit für bestimmte Konzentrationen bei der Ausbreitung fester Schadstoff-Teilchen in Industrie-Gebieten Staub-Reinhalt. Luft 33 (1973) 1, 20/24
Rouillard, E.E.A. Valvona, P.J.	Flow patterns upstream of isokinetic dust sampling probes Broschüre CSIR Report CENG 019 Feb. (1974)
Rüping, G.	Die Bedeutung der geschwindigkeitsgleichen Absaugung bei der Staubstrommessung mittels Entnahmesonden Staub-Reinhalt. Luft 28 (1968) 4, 137/144

Stepnewski, D.D., Smith, K.D. — Collect Samples Proportional to Varging Flow
Chem. Engng. 65 (1958) 14, 152/153

Umhauer, H. — Ermittlung von Partikelgrößenverteilungen in Aerosolströmungen hoher Konzentration mit Hilfe einer Streulichtmeßeinrichtung
Chem.-Ing.-Techn. 47 (1975) 7, 297 ff

Vitols, V. — Theor. Limits of Errors Due to Anisokinetic Sampling of Particulate Matter
J. Air Pollut. Contr. Ass. 16(1966)2, 79/83

Walkenhorst, W. — Feinste Siebe für die quantitative Probenahme von Schwebestoffen zur elektronenmikroskopischen Auswertung
Staub-Reinhalt. Luft 30 (1970) 6, 246/249

Watson, H.H. — Errors Due to Anisokinetic Sampling of Aerosols
Amer. Ind. Hyg. Assoc. Quart. 15 (1954) 21

Whiteley, A.B., Reed, L.E. — The effect of probe shape on the accuracy of sampling the gases for dust content
J. Inst. Fuel 32 (1959) 316/320

Sampling of special products

Badger, E.H.M. — Experiments on the sampling of coal; an account of recent work carried out for the British Standards Institution
J. Inst. of Fuel 26 (1953) 17/32, 59

Boreham, G.R. — Progress Review No. 44: Sampling, Analysis and Testing of Gaseous Fuels
J. Inst. Fuel 31 (1958) 210, 304/307

Coryell, R.L., Schwerd, F.J., Parente, E.J. — Test of accuracy of a mechanical coal sampler
Amer. Soc. Test. Mats. S.T.P. 162, 1954, 72/82

Davis, W., Hall, D.A. Tomlinson, R.C. — Methods of Testing Automatic Coal samples
J. Inst. Fuel 31 (1958) 212, 406/413

Emery, E.T.G. — The sampling of small coal
Engineering(London),Okt. 1951, 452/454

Flint, D., Ray, W.D. — Procedure for Coal Sample Preparation
J. Inst. Fuel 31 (1958) 206, 123/132

Gebhard, K. u.a. Untersuchungen über den Wirkungsgrad einer Erzbettungsanlage und die Probenahme von Eisenerzen
Stahl und Eisen 83 (1963) 15, 907/916

Hulley, B.J. Sampling and Sample Conditioning in online Fertilizer Analysis
Chem. Eng. Sci.? (1970)244,CE410/CE413

Islamov, M.S. Eine Probenahmeeinrichtung für Koks
Koks i. Chimija 1971, H. 5, 48/49

Landry, B.A., Anderson, W.W. Coal Sampling by large increment weights
Trans. Am. Soc. Mech. Engrs. 73 (1951), 989/996

Liddell, F.D.K. Coal Sampling
Fuel 30 (1951) 275

Lloyd, B.H. The statistical Approach to bulk Sampling Problems
Industrial Quality Control 9(1953)6,113

Miskimen, T.A., Thurston, R.S. Multilot sampling - the accuracies in sampling of large coal shipments by application of the variance concept
Amer. Soc. Test. Mats., S.T.P. No 162, 1954, 129/140

Sanada, M. Problems in gas sampling
Kagakukojyo (Chem. Factory)15(1971)4,77/84

Tanner, L. Probability Sampling Methods for Wool
Materials Research + Standards 1(1961)3,172

Tomlinson, R.C. The Routine Sampling of Coal
J. Inst. Fuel 26 (1953) 83

Tomlinson, R.C. Some recent British work on coal sampling
Am. Soc. Test. Mats-Spec. Techn. Publ. 162 (1954) 35/56

Subject Index

MIX
Papier aus verantwortungsvollen Quellen
Paper from responsible sources
FSC® C105338

If you have any concerns about our products,
you can contact us on
ProductSafety@springernature.com

In case Publisher is established outside the EU,
the EU authorized representative is:
Springer Nature Customer Service Center GmbH
Europaplatz 3, 69115 Heidelberg, Germany

Printed by Libri Plureos GmbH
in Hamburg, Germany